1 MONTH OF
FREE
READING

at

www.ForgottenBooks.com

By purchasing this book you are eligible for one month membership to ForgottenBooks.com, giving you unlimited access to our entire collection of over 700,000 titles via our web site and mobile apps.

To claim your free month visit:

www.forgottenbooks.com/free75345

ISBN 978-0-483-19281-2
PIBN 10075345

This book is a reproduction of an important historical work. Forgotten Books uses
state-of-the-art technology to digitally reconstruct the work, preserving the original format
whilst repairing imperfections present in the aged copy. In rare cases, an imperfection in
the original, such as a blemish or missing page, may be replicated in our edition. We do,
however, repair the vast majority of imperfections successfully; any imperfections that
remain are intentionally left to preserve the state of such historical works.

THE
NATURALIST'S
CABINET:
Containing
INTERESTING SKETCHES
OF
ANIMAL HISTORY;
Illustrative of the
NATURES, DISPOSITIONS, MANNERS, AND HABITS,
OF ALL THE MOST REMARKABLE
Quadrupeds, Birds, Fishes, Amphibia, Reptiles, &c.
IN THE KNOWN WORLD.

REGULARLY ARRANGED, AND ENRICHED WITH NUMEROUS
BEAUTIFUL DESCRIPTIVE ENGRAVINGS.

" Who can this field of miracles survey,
And not with *Galen* all in rapture say,
Behold a *God*, adore him, and obey'"

BLACKMORE.

IN SIX VOLUMES.
VOL. IV.

BY THE
REV. THOMAS SMITH,
Editor of a New and Improved Edition of Whiston's Josephus, &c. &c.

ALBION PRESS PRINTED:
PUBLISHED BY JAMES CUNDEE,
Ivy Lane, Paternoster-Row.

1806.

CONTENTS.

VOL. IV.

CONTENTS.

CONTENTS.

THE

𝕹aturalist's Cabinet.

CHAP. I.

" The cock, that is the trumpet of the morn,
Doth, with his lofty and shrill-sounding throat,
Awake the god of day ; and at his warning,
Whether in sea or fire, in earth or air,
Th' extravagant and guilty spirit hies
To his confine."

SHAKESPEARE.

THE COCK.

Introductory remarks.

THIS brave and majestic bird in his present state of domestication, differs so exceedingly from his wild original, as to render
it difficult to trace him back to his primitive
stock. According to the accounts of various
travellers, he is still found in a state of nature in
the forests of India, and in most of the islands
of the Indian seas. Sonnini asserts, that he has
seen wild cocks in the immense forests which
cover the interior of Guiana, which exhibited

A 2

peculiarities sufficient to prove they were natives of that climate, and not descended from birds of the same species brought from the ancient continent.

The appearance of the cock, when in his full plumage, is strikingly grand and animated. His head, which is small, is adorned with a beautiful red comb and wattles, his eyes sparkle with fire, and his whole demeanour bespeaks boldness and freedom. The feathers on his neck are long, and fall gracefully down upon his body, which is thick, firm, and compact. His tail is long, forming a beautiful arch, and giving grace to all his motions; his legs, which are strong, are armed with sharp spurs, with which he defends himself and attacks his antagonist. His whole aspect, when surrounded by his females, is full of animation; he allows of no competitor, but on the approach of a rival or an enemy, he rushes instantly to the combat, and either drives him from the field, or perishes in the attempt. In short, he remains entire master of the farm-yard, a representation of which scene, though familiar to our readers, accompanies this article.

" I have just witnessed," says M. de Buffon, " a singular scene. A sparrow-hawk alighted in a pretty populous court-yard; a young cock of this year's hatching darted at him and threw him on his back. In this situation the hawk defended himself with his talons and his bill, and intimidated the hens and the turkeys, which

screamed in tumultuous concert around him.
After he had a little recovered himself, he rose
and was taking wing, when the cock rushing
upon him a second time, overturned and held
him down so long that he was caught."

A similar instance occurred in January, 1804,
in the garden of Mr. Markwick, of Fittleworth,
in Sussex. A hawk pounced on a chicken about
half grown, and while encumbered with his prey
was perceived by the parent cock, who immediately made at the intruder, and by one blow laid
him at his feet. The chick was by this time disengaged, and a battle ensued between the hawk
and cock, which ended, after three rounds, in
favour of chanticleer, from whose feet Mr.
Markwick, who, with two other persons, had
witnessed the conflict, took the petty tyrant of
the air, with very few remaining symptoms of
life, which he presently resigned in his hands.

The cock is very attentive to his females,
scarcely ever losing sight of them. He leads,
defends, and cherishes them, collects them when
they straggle, and seems to eat unwillingly till
he sees them feeding around him, His affection
is only equalled by his jealousy; and the appearance of a strange cock in his domain is the signal
for an immediate battle. Nor is his jealousy entirely confined to his rivals; it is sometimes observed to extend even to his beloved female; and
he even appears capable of a certain degree of

of this remark, occurred not long ago, at the seat of a gentleman near Berwick, and is related by Dr. Percival, in his Dissertations. " My mowers," says this gentleman, " cut a partridge on her nest, and immediately brought the eggs, fourteen in number, to the house. I ordered them to be put under a very large beautiful hen, and her own to be taken away. They were hatched in two days, and the hen brought them up perfectly well till they were five or six weeks old. During that time they were constantly kept confined in an out-house, without having been seen by any of the other poultry. The door happening to be left open, the cock got in. My housekeeper hearing her hen in distress, ran to her assistance, but did not arrive in time to save her life. The cock, finding her with the brood of partridges, had fallen upon her with the utmost fury, and killed her. The housekeeper found him tearing her with both his beak and spurs; although she was then fluttering in the last agony, and incapable of any resistance. This hen had formerly been the cock's greatest favourite."

The varieties of this species are endless, almost every country producing a different kind. The principal selected for domestic purposes in this kingdom are the following:

The Hamburgh cock; this is a very large

kind, and much used for the table. It is also called velvet breeches, because his thighs and belly are of a soft black.

The bantam, or dwarf cock, is a diminutive, but very spirited breed. This species was first domesticated at Bantam, in India, (whence the name) and from thence transported into Europe. His legs are furnished with long feathers, which reach to the ground behind. He possesses great courage, and will fight with one much stronger than himself.

The frizzled cock has its feathers so curled up that they seem reversed and to stand in opposite directions. These are originally from the southern parts of Asia, and when young are extremely sensible of cold. They have a disordered and unpleasant appearance, but are in great esteem for the table.

The crested cock differs little in figure from the common cock, excepting in having a tuft of feathers on his head, and in general a smaller crest. The breed of crested cocks is that to which the curious have paid the greatest attention; they have consequently observed in it many differences, of which they have made just so many families, the individuals of which are the more highly esteemed in proportion to the beauty or rarity of their colours. Some of this species are destitute both of crest and wattle, and are in general long-legged. Sonnini says, that he has seen them of very large size, with thick long legs, a large tuft of feathers on the head, and plumage elegantly variegated, in the 'Thebaid, and particularly at Dendera. They are held in great estimation throughout all Egypt, on account of the excellent quality of their flesh.

The English game-cock is unrivalled by those of any other nation for his invincible courage, and is, on that account, employed as the instrument of the barbarous sport of cock-fighting. To trace this custom to its origin, we must look back to ancient times, and lament that it still continues the disgrace of a christian country.

When Themistocles, the celebrated Athenian commander, was marching to meet the Persians, who had invaded Greece, seeing that his soldiers manifested very little ardour to meet the enemy, he called their attention to the fury with which cocks engage each other. " Behold," said he, " the invincible courage of those animals; yet their only motive is their desire to conquer, while you fight for your homes, for the tombs of your fathers, for your liberty." This short address roused the latent courage of the army, and Themistocles obtained a signal victory. In honour of this event, the Athenians instituted a kind of festival, which was celebrated annually with cock-fights. From them the Romans are said to have learned the practice, and by that warlike people it was first introduced into this island. Henry VIII. was so fond of this sport that he erected a commodious house for the purpose, which, though now applied to a very different use, still retains the name of the cock-pit.

The following circumstance, that occurred in the month of April, 1789, proves how much this barbarous diversion tends to brutalize the mind: Mr. Ardesoif, of Tottenham, a young man of large fortune, was. passionately fond of cock-fighting. He possessed a favourite cock, which had won in many profitable matches; but losing, for once, his owner was so enraged that he ordered the bird to be tied to a spit and roasted alive before a large. fire. The screams of the miserable animal were so affecting that some gentlemen, who were present, attempted to interfere. This enraged him to such a degree that, seizing a poker, he declared with the most furious vehemence, he would kill the first person who should interpose. In the midst of these asseverations the inhuman perpetrator of this horrid deed fell down senseless on the spot, and upon being taken up was found to be dead. An instructive lesson for cock-fighters and gamesters of every description.

That all who delight in this cruel sport are not, however, always destitute of humanity may be inferred from the following recent anecdote of, a man in a more humble sphere of life. Nicholas Cannon, the driver of one of the Kentish stages, had a favourite game-cock, named Trumpeter, who had won every battle he ever fought,. but afterwards had the misfortune to break his leg in

a rat-trap. Cannon, who was uncommonly attached to this feathered hero, determined, if possible, to save his life; and striking off the broken
part of the limb, he gathered up the fibres of
the leg, and placed his favourite securely in a
sling, where he attended and fed him for five
weeks: he then took off the bandage, and found
the wound completely cicatrized. Possessing
considerable ingenuity, he next set about making
an artificial foot, and soon contrived a wooden
leg and foot, armed with a spur, and affixed it
to the stump of the amputated limb; upon this
the cock actually strutted among his barn-door
wives, at Canterbury, a terror to all his feathered
rivals.

In the island of Sumatra, the passion for cock-
fighting is so powerful, that the natives make it
a serious occupation rather than an amusement.
A man is seldom seen travelling in that country
without a cock under his arm; and, perhaps,
greater attention is paid by those people to the
rearing and breeding of these birds, than we ever
did when that diversion was at its height. They
arm one of the legs only, not with a slender gaff
as is the practice here, but with an instrument in
the form of a scymetar, with which the animals
make most terrible destruction. The Sumatrians
fight their cocks for vast sums: a man has been
known to stake his wife, or daughter, and a son
his mother, or sisters, on the issue of a battle.
Four arbitrators are appointed to decide in all

disputed points, and if they cannot agree there
is no appeal but to the sword. Some of them
have a notion that their cocks are invulnerable;
under this persuasion a father has been known,
on his death-bed, to direct his son to lay his
whole property on a certain bird, as if confident
of success.

The fecundity of the hen is great; she generally lays two eggs in three days, and continues
to lay through the greatest part of the year, excepting the time of moulting, which lasts about
two months. After laying about twenty-five or
thirty eggs, she prepares for the painful task of
incubation, in which her patience and perseve-
rance are truly extraordinary. A sitting hen is
a lively emblem of the most affectionate solici-
tude and attention; she covers her eggs with
her wings, fosters them with a genial warmth,
changing them gently that all the parts may receive an equal degree of heat. She seems perfectly sensible of the importance of her employment, and is so intent on her occupation as to
neglect in some measure the necessary supplies
of food and drink. In about three weeks the
young brood burst from their confinement, and
the hen, from the most cowardly and voracious,
becomes, in the protection of her young, the
most daring and abstemious of animals.

As the chickens reared by the hen bear no
proportion to the number of eggs she produces,
many artificial schemes of rearing have been at- .

tempted. The most successful, though by no
means the most humane, is said to · be where a
capon is made to supply the place of a hen. He
is rendered very tame; the feathers are plucked
from his breast, and the bare parts are rubbed
with nettles. The chickens are then put to him;
and by their running under his breast with their
soft and downy bodies, his pain is so much
allayed, and he feels so much comfort to his fea-
therless part, that he soon adopts them, feeding
them like a hen, and assiduously performing all
the functions of the tenderest parent.

Chickens have long been hatched in Egypt by
means of artificial heat. This is now principally
practised by the inhabitants of a village called
Berme, and by those who live at a little distance
from it. Towards the beginning of autumn,
these persons spread themselves all over the
country; and each of them is ready to undertake
the management of an oven. These ovens are
of different sizes, each capable of containing
from forty to eighty thousand eggs; and the
number of ovens in different parts is about three
hundred and eighty-six. They are usually kept
in exercise for about six months; and as each
brood takes up twenty-one days in hatching, it
is easy in every one of them to produce eight
different broods of chickens in the year. The
ovens where these eggs are placed, are of the most
simple construction; consisting only of a low
arched apartment of clay. Two rows of shelves

are formed, and the eggs are placed on these in such a manner as not to touch each other. They are slightly moved five or six times in every twenty-four hours. All possible care is taken to diffuse the heat equally throughout; and there is but one aperture, just large enough to admit a man stooping. During the first eight days the heat is rendered great; but during the last eight it is gradually diminished, till at length, when the young brood are ready to come forth, it is reduced almost to the state of the natural atmosphere. At the end of the first eight days it is known which of the eggs will be productive.

Every person who undertakes the care of an oven, is under the obligation only of delivering to his employer two-thirds of as many chickens as there have been eggs given to him; and he is a gainer by this bargain, as it always happens, except from some unlucky accident, that many more than that proportion of the eggs produce chickens. It is calculated that the ovens in Egypt annually give life to almost a hundred millions of these animals.

The ingenious M. de Reaumur introduced this useful and advantageous mode of hatching eggs into France. By a number of experiments, he reduced the art to certain principles. He found that the degree of heat necessary for producing all kinds of domestic fowls was the same; the only difference consisted in the time during which it ought to be communicated to the eggs: it will bring the canary-bird to perfection in eleven or twelve days, while the turkey-poult requires twenty, or twenty-eight. He also found that stoves, heated by means of pipes from a baker's oven, or the furnaces of glass-houses, succeeded better than those made hot by the layers of dung, the mode preferred in Egypt. These should have their heat kept as nearly equal as possible; and the eggs should be frequently removed from the sides into the middle, in order that each may receive an equal portion. After his eggs were hatched, he had the offspring put into a kind of low boxes, without bottoms, and lined with fur; whose warmth supplied the place of a hen, and in which the chickens could at any time take shelter. These were kept in a warm room till the chickens acquired some strength; they then could be placed, with safety, exposed to the open air, in a court-yard. The young brood are generally a whole day, after being hatched, before they take any food at all; and then a few crumbs of bread are given for a day or two, after which time they begin to pick up insects and grains for themselves. But, in order to save the trouble of attending them, capons are taught to watch them, the same as hens. M. de Reaumur says, that he has seen above two hundred chickens at once, all led about and defended by only three or four capons. It is asserted, that even cocks may be taught to perform this office; which they will continue to do all their lives afterward.

Eggs hatched by the heat of the human body.

The heat communicated by the human body has been also found capable of hatching eggs Livia, a Roman lady of distinction, being pregnant, took a fancy to hatch an egg in her bosom, with a view to augur the sex of her child from that of the chicken it produced. The chicken was a male, and so was the child. What Livia executed, to gratify her curiosity, was undertaken, in the year 1706, by a young lady of Barre, in order to procure a treat for her tutor. This young lady, whose good sense, virtue, and piety, are highly extolled in the Clef du Cabinet, in which this fact is related, pretended to be so ill as to be obliged to keep her bed. During this time she endeavoured, by the heat of her body, to hatch a turkey's egg; she was successful, and took particular care of the bird, which owed both its life and death to her, till it attained the weight of seven pounds, when she dressed it; and, (continues the above-mentioned journal), the man, whom she was desirous to please, acknowledged that he had never tasted any thing so exquisitely delicate.

The following experiment, of which we find an account in Buffon's history, is of a nature too singular to be omitted. It consists in cutting off the crests of a pullet, and inserting in the place one of the spurs, which are then just beginning to spring forth. The spurs, grafted in this manner, by degrees take root in the flesh, derive nourishment from it, and frequently grow to a greater length than they would have done in their proper place. They have been seen two inches and a half long, and, three lines and a half in diameter at the base; sometimes they grow curved like ram's horns, and at others like those of goats.

The progress of the incubation of the chicken, in the natural way, is a subject too curious and too interesting to be passed over without notice. The hen has scarcely sat on the eggs twelve hours, when some lineaments of the head and body of the chicken appear. The heart may be seen to beat at the end of the second day: it has at that time somewhat the form of a horse-shoe, but no blood yet appears. At the end of two days two vesicles of blood are to be distinguished, the pulsation of which is very visible: one of these is the left ventricle, and the other the root of the great artery. At the fiftieth hour, one auricle of the heart appears, resembling a noose folded down upon itself. The beating of the heart is first observed in the auricle, and afterwards in the ventricle. At the end of seventy hours, the wings are distinguishable; and on the head two bubbles are seen for the brain, one for the bill, and two others for the fore and hind part of the head. Towards the end of the fourth day, the two auricles, already visible, draw nearer to the heart than before. The liver appears towards the fifth day. At the end of a hundred and thirty-one hours, the first voluntary motion is observed. At the end of seven hours more, the lungs and stomach become visible; and, four hours after this, the intestines, the loins, and the upper jaw. At the hundred and forty-fourth hour, two ventricles are visible, and two drops of blood, instead of the single one which was seen before. The seventh day, the brain begins to have some consistence. At the hundred and ninetieth hour of incubation, the bill opens, and the flesh appears in the breast; in four hours more, the breast-bone is seen; and in six hours after this, the ribs appear forming from the back, and the bill is very visible, as well as the gall-bladder. The bill becomes green at the end of two hundred and thirty-six hours; and if the chicken is taken out of its coverings, it evidently moves itself. The feathers begin to shoot out towards the two hundred and fortieth hour, and the skull becomes gristly. At the two hundred and sixty-fourth hour, the eyes appear. At the two hundred and eighty-eighth, the ribs are perfect. At the three hundred and thirty-first, the spleen draws near the stomach, and the lungs to the chest. At the end of three hundred and fifty-five hours, the bill frequently opens and shuts; and at the end of the eighteenth day, the first cry of the chicken is heard. It afterwards gets more strength, and grows continually, till at length it is enabled to set itself free from its confinement.

During this whole process, every part appears

liver is formed on the fifth day, it is founded on
the preceding situation of the chicken, and on
the changes that were to follow.   No part of the
body could possibly appear either sooner or later,
without the whole embryo suffering; and each
of the limbs becomes visible at the fit moment.
This ordination, so wise, and so invariable, is
manifestly the work of a Supreme Being: but
we must still more sensibly acknowledge his
creative powers, when we consider the manner in
which the chicken is formed out of the parts
which compose the egg.   How astonishing must
it appear, to an observing mind, that in this substance there should be, at all, the vital principle,
of an animated being.   That all the parts of an
animal's body should be concealed in it, and require nothing but heat to unfold and quicken
them.   That the whole formation of the chicken
should be so constant and regular.   That, exactly at the same time, the same changes will

chicken, the moment it is hatched, is heavier
than the egg was before !

THE WOOD GROUS, OR COCK OF THE WOOD

IS almost the size of a turkey, and often weighs near fourteen pounds; but the female is much smaller. The head and neck are ash-colour, crossed with black lines; the body and wings chesnut brown, and the breast of a very glossy blackish green. The legs are strong, and cover-ed with brown feathers. The plumage of the female differs from this description, in being red about the throat, and having the head, neck, and back, crossed with red and black bars; the belly barred with orange and black, with the tips of the feathers white, as are also the tips of the shoulders; indeed, she is altogether so very different, that she might be supposed to belong to another species.

The wood grous is chiefly fond of a mountain-ous, or woody situation. In winter he resides in the deepest recesses of the woods, and in sum-mer he ventures down from his seclusion, to make short depredations on the farmer's corn: but in these excursions he seems to be perfectly aware of his danger, and is constantly upon his guard; so much so, indeed, that it is then very difficult to come near him by surprise; and very few are taken but by those who in autumn pur-sue him into his natural retreats, and which is often done from his flesh being considered as very delicate food.

When in the forest, the wood grous attaches himself principally to the oak and the pine-tree; the cones of the latter serving for his food, and the thick boughs for an habitation; and be sometimes will strip one tree bare before he attempts the cones of another. He feeds also upon ants' eggs, which seem a high delicacy to all birds of the poultry kind; cranberries are likewise found in his crop; and his gizzard, like that of domestic fowls, contains a quantity of gravel, for the purpose of assisting his powers of digestion.

This bird begins to feel the genial influence of the spring at its first approach, and his season of love may be said to continue from that time until the trees have all their leaves, and the forest is in full bloom. During this whole season he may be seen at sun-rise and setting, extremely active upon one of the largest branches of the pine-tree. With his tail raised and, expanded like a fan, and the wings drooping, he walks backward and forward, his neck stretched out, his head swollen and red, and making a thousand ridiculous postures: his cry upon that occasion, is a kind of loud explosion, which is instantly followed by a noise like the whetting of a scythe, which ceases and commences alternately for about an hour, and is then terminated by the same explosion. During the time he continues this

singular cry he seems entirely deaf, and insensible
of every danger: whatever noise may be made
near him, or even though fired at, he still unconcernedly continues his call. Upon all other occasions he is the most timid and watchful bird in
nature: but then he seems entirely absorbed by
his instincts, and seldom leaves the place where
he first begins to express the excesses of desire.
This extraordinary cry, which he accompanies
by a clapping of the wings, is no sooner finished
than the females who hear it, reply, approach,
and place themselves under the tree, from whence
the male descends to them. The number that, on
this occasion, resort to his call, is uncertain; but
one male generally suffices for all the females in
one part of the forest. The female seldom lays
more than six or seven eggs, which are white, and
marked with yellow, of the size of a common hen's
egg: she generally lays them in a dry place, and
a mossy ground, and hatches them without the
company of the male. When she is obliged,
during the time of incubation, to leave her eggs
in quest of food, she covers them up so artfully
with moss, or dry leaves, that it is extremely difficult to discover them; and when sitting, though
wild and timorous at other times, she will suffer
the sportsmen to approach and drag her off
her nest. She often keeps to her nest though
strangers attempt to drag her away.

As soon as the young ones are hatched they
run with extreme agility after the mother, some

times even before they are entirely disengaged from the shell. The hen leads them forward to procure ants' eggs, and the wild mountain-berries, which while young are their only food. As they grow older, they feed upon the tops of heath; and the cones of the pine-tree. In this manner they soon come to perfection; they are an hardy bird, their food lies every where before them, and it would seem that they should increase in great abundance; but this is not the case; their numbers are thinned by rapacious birds and beasts of every kind, and still more by their own salacious contests.

The whole brood follows the mother for about two months, at the end of which the young males entirely forsake her, and keep in great harmony together till the beginning of spring, when they bid adieu to all their former amity. They then consider each other as rivals, fight like game-cocks, and are so inattentive to their own safety that it often happens that two or three of them are killed at a shot.

THE RUFFED GROUS

IS in size between that of a pheasant and a partridge. The bill is brownish. The head is crested; and, as well as all the upper parts, is variegated with different tints of brown mixed with black. The feathers on the neck are long and loose; and may be erected at pleasure, like

those of the cock. · The throat and the fore part
of the neck are orange brown; and the rest of the
under parts yellowish white, having a few curved
marks on the breast and sides. The tail consists
of eighteen feathers; all of which are crossed with
narrow bars of black, and one broad band of the
same near the end. The legs are covered to the
toes (which are flesh-coloured, and pectinated on
the sides) with whitish hairs.

The Ruffed Grous which has hitherto been
found only on the new continent, is a fine bird
when he displays his gaiety, spreading his tail like
that of a turkey-cock, and erecting the circle of
feathers round his neck like a ruff; walking very
stately with an even pace, and making a noise
somewhat like a turkey. This is the moment that
the hunter seizes to fire at him; for if the bird
sees that it is discovered, it immediately flies off
to the distance of some hundred yards before it
again settles.

The *thumping,* as it is called, of these birds is
very remarkable. This they do, by clapping their
wings against their sides. They stand upon an
old fallen tree, that has lain many years on the
ground; in which station they begin their strokes
gradually, at about two seconds of time from one
another, and repeat them quicker and quicker until they make a noise not unlike distant thunder.
This continues from the beginning about a minute; the bird ceases for six or eight minutes,
and then begins again. The sound is often heard

3

at the distance of nearly half a mile ; and sportsmen take advantage of this note, to discover the
birds, and shoot them. The Grous commonly
practise their *thumping* during the spring and fall
of the year ; at about nine or ten o'clock in the
morning, and four or five in the afternoon.

The females lay their eggs, from twelve to sixteen in number, in nests which they make either
by the side of fallen trees, or the roots of standing
ones. Mr. Brook of Maryland, in North America, speaking of this bird, says " I have found
their nests when a boy; and have endeavoured
to take the old bird, but never could succeed : she
would let me put my hand almost upon her before she would quit her nest ; then by artifice she
would draw me off from her eggs, by fluttering
just before me for a hundred paces or more, so
that I have been in constant hopes of taking her
When the nestlings are hatched, and a few days
old, they hide themselves so artfully among the
leaves, that it is difficult to find them."

THE BLACK GROUS.

THE name of this bird almost furnishes a description, the whole body being black, but it has
another remarkable characteristic, which is that
its tail is forked. The weight of the male is about
four pounds, and that of the female about two.
These birds were formerly to be found in great

abundance in the north of England, but they have now become very scarce. This is owing to various causes; but principally to the great improvement in the art of shooting-flying, and to the inclosure of waste lands. Some few are yet found in Wales; and in particular parts of the New Forest in Hampshire they are in tolerable plenty, being preserved as royal game, and always excepted in the warrants to kill game there. They are partial to mountainous and woody situations, far removed from the habitations of men.

Their food is various; but principally consists of the mountain fruits and berries, and in winter the tops of heath. It is somewhat remarkable that cherries and pease are fatal to these birds. They perch and roost in the same manner as the pheasant. They never pair; but in the spring the males assemble at their accustomed resorts on the tops of heathy mountains when they crow and clap their wings, like the wood grous.

The female forms an artless nest on the ground; and lays six or eight eggs of a dull yellowish white colour, marked with numerous very small ferruginous specks, and towards the smaller end with some blotches of the same. These are hatched very late in the summer. The young males quit their parent in the beginning of winter, and keep together in flocks of seven or eight till the spring.

These birds will live and thrive in menageries, but they have not been known to breed in a state of confinement. In Sweden, however, a spurious breed has sometimes been produced with the domestic hen.

In Russia, Norway, and other extreme northern countries, the black grous are said to retire under the snow during winter. The shooting of them in Russia is thus conducted. Huts full of loop-holes, like little forts, are built for this purpose, in woods frequented by these birds. Upon the trees within shot of the huts, are placed artificial decoy birds. As the grous assemble, the company fire through the openings; and so long as the sportsmen are concealed, the report of the guns does not frighten the birds away. Several of them may therefore be killed from the same tree, when three or four happen to be perched on branches one above another. The sportsman has only to shoot the undermost bird first, and the others upward in succession. The uppermost bird is earnestly employed in looking down after his fallen companions, and keeps chattering to them till he becomes the next victim.

The inhabitants of Siberia, during winter, take these birds in the following manner. A number of poles are laid horizontally on forked sticks, in the open birch forests. Small bundles of corn are tied on these, by way of allurement; and at a ittle distance some tall baskets of a conical shape are placed, having their broad part uppermost. Just within the mouth of each basket, is placed a small wheel; through which passes an axis so cely fixed, as to admit it to play very readily,

Description—Habits.

and on the least touch either on one side or the other to drop down and again recover its situation. The black grous are soon attracted by the corn on the horizontal poles. The first comers alight upon them, and after a short repast, fly to the baskets, and attempt to settle on their tops, when the wheel drops sideways, and they fall headlong into the trap. These baskets are sometimes found half-full of birds thus caught.

THE RED GROUS.

THIS species is rather smaller than the preceding, the weight of the male being about nineteen, and that of the female fifteen ounces. These birds abound in the healthy and mountainous parts of the northern counties of England. They are likewise very common in Wales, and the Highlands of Scotland; but they have not yet been observed in any of the countries of the continent.

In winter they are usually found in flocks of sometimes forty or fifty, which are termed by sportsmen, ' packs'; and become remarkably shy and wild. They keep near the summits of the heathy hills, seldom descending to the lower grounds. Here they feed on the mountain berries, and on the tender tops of the heath.

They pair in spring; and the females lay from six to ten eggs, in a rude nest formed on the ground. The young brood (which during the first year are called poults) follow the hen till the approach of winter; when they unite with several others into packs.

Red Grous have been known to breed in confinement, in the menagerie of the late Duchess Dowager of Portland. This was in some measure effected by her Grace causing fresh pots of heath to be placed in the menagerie almost every day.

The flesh, of the red grous, as well as of all the other species, is an excellent food, but very soon corrupts. To prevent this, Mr. Daniel and other sporting writers say the birds should be drawn immediately after they are shot. Among the varieties of this species are the hazle grous (a native of Germany) and the pin-tailed grous, so called from its narrow forked tail.

THE PARTRIDGE.

THE length of this bird is about thirteen inches; on the breast it has a crescent of a deep chesnut colour, and under each eye there is a small coloured spot, which has a granulated appearance, and extends behind the eye. The sides of the head are yellowish, and the general colour of the plumage is brown and ash, elegantly mixed with black. The wings are brown with dark bars; the tail is short, and consists of eighteen

Description—Manners.

feathers, of which the seven nearest the sides are red, with an ash-coloured border. Sportsmen as well as naturalists have believed that the female has no crescent on the breast, like the male. This, however, on dissection has proved to be a mistake; for Mr. Montague happening to kill nine birds in one day, with very little variation as to the mark on the breast, was led to open them all, and discovered that five of them were females. On carefully examining the plumage, he found that the males could only be known by the superior brightness of colour about the head which alone after the first or second year seems to be the mark of distinction.

Partridges are found principally in temperate climates: they are no where in greater abundance than in this island, and form a part of the most elegant entertainments. They pair early in the spring; the female makes a nest of grass and dry leaves on the ground, and lays from fifteen to twenty and sometimes five and twenty eggs. Mr. Daniel says thirty-three eggs have been found in one nest and of these twenty-three produced young ones. They run the moment they are hatched, frequently carrying along with them part of their shell.

The male shares with his mate the trouble of rearing their young. They immediately lead them to ant-hills, on the grubs of which insects they at first principally feed: for, at this season the various species of ants loosen the earth about their habitations. The young birds therefore have only to scrape away the earth, and they can satisfy their hunger without difficulty. A covey that some years ago invited the attention of the Rev. Mr. Gould, gave him an opportunity of remarking the great delight they take in this kind

withdrawing to some distance, the parent birds

heartily. After a few days, they grew more bold, and ventured to eat within twelve or fourteen yards of him. The surrounding grass was high; by which means they could, on the least disturbance, immediately run out of sight, and conceal themselves. The excellence of this food for partridges may be ascertained from those that are bred up under a domestic hen, if constantly supplied with ants' grubs and fresh water, seldom failing to arrive at maturity. Along with the grubs it is recommended to give them, at intervals, a mixture of millepedes, or wood-lice, and earwigs to prevent their surfeiting on one luxurious diet; fresh curds mixed with lettuce, chickweed, or groundsel should also be given them.

The parents frequently sit close by each other, covering their young brood with their wings. In this situation they are not easily flushed; and a sportsman, attentive to the preservation of his game, will avoid disturbing them in a performance of a duty so truly interesting. If, however, a dog should approach too near, the male always

runs off first with a peculiar cry of distress; he
stops at the distance of thirty or forty paces, and
frequently returns several times towards the dog,
clapping his wings; with such courage does paternal affection inspire even the most timid of
animals. He then flies, or rather runs, heavily
along the ground, dragging his wings, as if to allure the enemy by the hope of an easy prey, making off fast enough to avoid being taken, and yet
so slow as not to discourage his pursuer till he
has at length decoyed him to a considerable distance from the covey. The female flies away to a
greater distance, and in a different direction; but
immediately returns, running along the ground
and finds her brood squatted among the grass
and leaves. Calling them hastily together, she
leads them, unperceived by the sportsman, to a
great distance, before the dog has time to return
from the pursuit of the male.

This bird flourishes best in cultivated countries, living principally on the labors of the husbandman : the extremes of heat and cold, are un-
favourable to its propagation. Mr. White, who
gives an instance of its instinctive sagacity; informs us that " a partridge came out of a ditch,
and ran along shivering with her wings and crying
out as if wounded and unable to get from us.
While the dam feigned this distress, a boy who
attended me, saw the brood, which was small
and unable to fly, run for shelter into an old fox's
hole, under the bank."—Mr. Markwick also re-

lates that " as he was once hunting with a young pointer, the dog ran on a brood of very small partridges. The old bird cried, fluttered, and ran tumbling along Just before the dog's nose, till she had drawn him to a considerable distance; when she took wing and flew farther off, but not out of the field. On this the dog returned nearly to the place where the young ones lay concealed in the grass; which the old bird no sooner perceived,

dog's nose, and a second time acted the same part, rolling and tumbling about till she drew off his attention from her brood, and thus succeeded in preserving them."—This gentleman says also, that when a kite was once hovering over a covey of young partridges, he saw the old birds fly up at the ferocious enemy, screaming and fighting with all their might to preserve their brood.

The eggs of the partridge are frequently destroyed by weesels, stoats, crows, magpies, and other animals. When this has been the case, the female frequently makes another nest and lays afresh. The produce of these second hatchings are those small birds that are not perfectly

This is always a puny, sickly race; and the individuals seldom outlive the rigours of the winter.

Those partridges which are hatched under a domestic hen, are said to retain through life the habit of *calling* whenever they hear the clucking of hens.

Instance of one remaining tame.

This bird, even when reared by the hand, soon neglects those who have the care of it; and shortly after its full growth, altogether estranges itself from the house where it was bred. This will invariably be its conduct, however intimately it may have connected itself with the place and inhabitants in the early part of its existence. Among the very few instances of the partridge's remaining tame, was that of one reared by the Rev. Mr. Bird, as recorded by Mr. Daniel. This, long after its full growth, attended the parlour at breakfast and other times, received food from any hand that gave it, and stretched itself before the fire and seemed much to enjoy the warmth. At length, it fell a victim to the decided foe of all favourite birds, a cat.

The same author also informs us that on the farm of Lion Hall, in Essex, belonging to Colonel Hawker, a partridge, in the year 1788, formed her nest, and hatched sixteen eggs, *on the top of a pollard oak tree*. What renders this circumstance the more remarkable is, that the tree had, fastened to it, the bars of a stile, where there was a footpath; and the passengers in going over, discovered and disturbed her before she sat close. When the brood was hatched they scrambled down the short and rough boughs, which grew out all around from the trunk of the tree, and reached the ground in safety.

The following occurrence took place at East Dean in Sussex in 1798; which will tend to prove

that partridges have no powers of migration. A
covey of sixteen partridges being routed by some
men at plough, directed their flight across the
cliff to the sea, over which they continued their
course about three hundred yards. Either intimidated or otherwise affected by that element,
the whole were then observed to drop into the
water. Twelve of them were soon afterwards
floated to shore by the tide; where they were
picked up by a boy, who carried them to East-
bourne and sold them.

Willoughby, as a proof of the docility of partridges, informs us, that a certain Sussex man
had, by his industry, made a covey of these birds

wager, out of the above mentioned county to
London, though they were absolutely free, and
-had their wings grown. An engraved representation of this singular fact is herewith presented to our readers.

In Sweden, these birds burrow beneath the
snow, and the whole covey crouds together under
shelter to guard against the intense cold. In
Greenland, the partridge is brown during summer,
but as soon as the winter sets in, it becomes
clothed with a thick and warm down, and its exterior assumes the colour of the snows. Near the
mouth of the river Oi in Russia, the partridges
are in such quantities, that the adjacent mountains are crouded with them. These birds have
been seen variegated with white, and sometimes

entirely white, where the climate could not be supposed to have any influence in this variation, and even among those whose plumage was of the usual colour.

THE TURKEY.

IT is generally believed that this bird is a native of North America, and was introduced from thence into England in the reign of Henry the Eighth. According to Tusser's, "Five Hundred Pointes of good Husbandrie," it began about the year 1585 to form an article in our rural Christmas feasts. It is a large, but unweildy bird; the anterior part of the head is strangely covered and ornamented with a pendulous, soft, fleshy substance; as are all the sides of the head and throat: the eyes are small, but bright and piercing; the bill convex, short, and strong; a long tuft of coarse black hairs on the breast; the wings moderately long, but not at all formed for supporting so large a bulk in long flights; the legs moderately long, and very robust. The plumage is dark, glossed with variable copper and green; the coverts of the wings and the quill-feathers barred with black and white. The tail consists of two orders; the upper, or shorter, very elegant; the ground colour a bright bay, the middle feathers marked with numerous bars of shining black and green. The longer, or lower order, is of a rusty white colour, mottled with black, and crossed with numerous narrow-waved lines of the same colour, and near the end with a broad band.

The turkey is one of the most difficult birds to rear of any that we have; and yet, in its wild state, it is found in great plenty, in the forests of Canada, that are covered with snow above three-fourths of the year. They are particularly fond of the seeds of nettles; but the seeds of the fox glove are a deadly poison to them.

The hunting of these birds forms one of the principal diversions of the natives of Canada. When they have discovered the retreat of the turkies, which in general is near fields of nettles, or where there is plenty of any kind of grain, they send a well-trained dog into the midst of the flock. The birds no sooner perceive their enemy, than they run off at full speed, and with such swiftness, that they leave the dog far behind. He, however, follows; and, as they cannot go at this rate for any length of time, at last forces them to take shelter in a tree : where they sit, perfectly spent and fatigued, till the hunters come up, and with long poles knock them down, one after another.

Turkies are among themselves extremely furious, and yet against other animals they are generally weak and cowardly. The domestic cock often makes them keep at a distance; and the latter seldom venture to attack him but with

united force, when the cock is rather oppressed
by their weight, than annoyed by their weapons.
There have, however, occurred instances in which
the turkey-cock has not been found wanting in
prowess :—A gentleman of New York received
from a distance a turkey-cock and hen, and a pair
of bantams, which he put into his yard, with
other poultry. Some time after, as he was feeding them from the barn-door, a large hawk suddenly turned the corner of the barn, and made a
pitch at the bantam hen: she immediately gave
the alarm, by a noise which is natural to her on
such occasions ; when the turkey-cock, who was
at the distance of about two yards, and, no doubt,
understood the hawk's intentions, and the immi-
nent danger of his old acquaintance, flew at the
tyrant with such violence, and gave him so severe a stroke with his spurs, when about to seize
his prey, as to knock him from the hen to a considerable distance ; and the timely aid of this
faithful auxiliary completely saved the bantam
from being devoured.

Another anecdote (though very different in its
nature) is recorded of the gallantry of the turkey-
cock; which also affords a singular example of
deviation from instinct. In May, 1798, a female turkey, belonging to a gentleman in Sweden, was sitting upon eggs ; and as the cock, in
her absence, began to appear uneasy and deject-
ed, he was put into the place with her. He immediately sat down by her side ; and it was soon

found that he had taken some eggs from under her, which he covered very carefully. The eggs were put back, but he soon afterwards took them' again. This induced the owner, by way of ex-periment, to have a nest made, and as many eggs put in as it was thought the cock could con-veniently cover. The bird seemed highly pleased with this mark of confidence; he sat with great patience on the eggs, and was so attentive to the care of hatching them, as scarcely to afford him-self time to take the food necessary for his sup-port. At the usual period, twenty-eight young ones were produced; and the cock, who was, in some measure; the parent of this numerous off-spring, appeared perplexed on seeing so many little creatures picking around him, and requir-ing his care. It was, however, thought proper not to entrust him with the rearing of the brood, lest he should neglect them; they were, there-fore, taken away and reared by other means.

. The female is, in general, much more mild and gentle than the male. When leading out her young family to collect their food, though so large, and apparently so powerful a bird, she gives them very little protection against the at-tacks of any rapacious animal that comes in her way. She rather warns them to shift for them-selves, than prepares to defend them. " I have heard a turkey-hen, when at the head of her brood, (says the Abbe de la Pluche) send forth the most hideous scream, without my being able to perceive the cause: her young, however, immediately when the warning was given, skulked under the bushes, the grass, or whatever else seemed to offer shelter or protection. They even stretched themselves at their full length on the ground, and continued lying motionless, as if dead. In the mean time the mother, with her eyes directed upwards, continued her cries and screaming, as before. On looking up, in the direction in which she seemed to gaze, I discovered a black spot just under the clouds, but was unable, at first, to determine what it was: however, it soon appeared to be a bird of prey, though at first at too great a distance to be distinguished. I have seen one of these animals continue in this agitated state, and her whole brood pinned down as it were to the ground, for four hours together; whilst their formidable foe has taken his circuits, has mounted, and hovered directly over their heads: at last, upon his disappearing, the parent changed her note, and sent forth another cry, which, in an instant, gave life to the whole trembling tribe, and they all flocked round her with expressions of pleasure, as if conscious of their happy escape from danger." As language can give but a faint description of this maternal agitation, our artist's pencil has been employed, to express it more fully.

In the wilds of America the turkey grows to a much larger size than with us. Josselyn says,

that he has eaten part of a turkey-cock, which,
after it was plucked, and the entrails were taken
out, weighed thirty pounds. Lawson saw half a
turkey serve eight hungry men for two meals,
and says that he has seen others which he believed would each weigh forty pounds. Some
writers even assert, that instances have occurred
of Turkies weighing no less than sixty pounds.

The females lay their eggs in spring, generally
in some retired and obscure place ; for the cock,
enraged at the loss of his mate, while she is employed in hatching, is apt otherwise to break
them. They sit on their eggs with so much per-
severance, that, if not taken away, they will almost perish with hunger before they will entirely
leave the nest. They are exceedingly affectionate to their young.

These birds are bred in great numbers in Norfolk, Suffolk, and some other counties, from
whence they are driven to the London markets,
in flocks of several hundreds. The drivers manage them with great facility, by means of a bit
of red rag tied to the end of a long stick ; which,
from the antipathy these birds bear to that
colour, effectually answers the purpose of a
scourge. ·

Notwithstanding the difficulty of rearing these
birds, yet the famous Bisset (as mentioned in the
" Eccentric Mirror,") taught six turkey-cocks to
go through a regular country-dance; but, in do-

ing this, he confessed he adopted the eastern method of heating the floor, by which camels are taught to dance. ,   . :   · ·, bl r!, · ·

Turkies, in a wild state, are gregarious; and associate in flocks, sometimes of five hundred. They frequent the great swamps of America to roost; but leave these situations at sun-rise, to repair to the dry woods, in search of acorns and berries. They perch on trees, and gain the height they wish by rising from bough to bough : they generally mount to the summits of even the loftiest, so as to be often beyond musket-shot. They are very swift runners, but fly awkwardly; and, about the month of March, they become so fat that they cannot fly beyond three or four hundred yards, and are then easily run down by a horseman.

Wild turkies are now very seldom seen in the inhabited parts of America: they are only found, in any great numbers, in the distant and most unfrequented parts. If the eggs of wild turkies be hatched under tame ones, the young are said still to retain a certain degree of wildness, and to perch separate from the others: yet they will mix and breed together in the season. The Indians sometimes use the breed produced from the wild birds, to decoy within their reach those still in a state of nature. They also make an elegant clothing of the feathers. They twist the inner webs into a strong double string, with hemp, or

Utility.

the inner bark of the mulberry-tree, and work it
like matting.   This appears very rich and glossy,
and as fine as silk shag.   The natives of Louisiana make fans of the tail; and, of four tails
joined together, the French used formerly to con-
struct a parasol.

CHAP. II.

The *Swallow* oft beneath my thatch
  Shall twitter from her clay-built nest,·
Oft shall the pilgrim lift the latch,
  And share my meal, a welcome guest.

ROGERS.

THE SWALLOW.

THERE are said to be upwards of six and thirty distinct species of this tribe, but only seven
or eight are known in these climates. The common house or chimney swallow in general measures about six inches from the point of the bill
to the end of the tail, and from the tip of the
one wing to the tip of the other, when extended,
seldom less than eleven or twelve; and commonly
weigh from fourteen to fifteen drachms. They
have a short black bill, but very broad at the
base, so that they are enabled to open their
mouths very wide; they have large eyes of an
hazel colour; the head, neck, and upper parts of
the body, are of a fine shining purplish blue,
with an orange-coloured spot above the bill, and
another of the same colour underneath; the breast

F 2

and belly are of a dusky white, with a slight reddish shade: the covert feathers are of the same colour of the back and head, but the quill-feathers are of a perfect black; they have a pretty long tail, made up of twelve feathers, which' is forked, the outermost feathers being pointed and near an inch longer than the others; they are all black, except the two middlemost, with a white spot upon each, which spot makes a beautiful line which crosses the tail, interrupted only by the two middle feathers.

Swallows are easily distinguished from all other birds, not only by their structure, but by their twittering voices and their manner of life. They fly with great rapidity, seldom walk, and perform all their functions either on the wing or sitting. They feed chiefly upon flies, gnats, and beetles, which they catch mostly, if not entirely, while flying; and it is for this purpose that they are continually hovering over pools and standing waters, where insects frequent without number. They often settle on dug ground or paths for gravel, which assists in grinding and digesting their food. They have an exceeding sharp eye, and are so fleet and uncertain in their course, that it is no uncommon thing for a sportsman to declare war against, and seek the constant destruction of these harmless animals, merely for the purpose of rendering himself a good shot. All the tribe drink as they fly along, sipping the surface of the water, but the house swallow

alone, in general, washes on the wing, by dropping into a pool several times successively.

It was a long time a matter of doubt, among
the most eminent naturalists, whether these birds
went into a state of torpor in the winter, or emi-
grated into other countries; in support of the
former opinion were the Hon. Daines Barring.
ton, and other celebrated writers. Dr. Fry asserted, that he was told by a fisherman, who was
accounted a man of veracity, that being near
some rocks on the coast of Cornwall, he saw, at
a very low ebb, a black list of something adhering to the rock, which, when he came to exa-
mine, he found it was a great number of swallows and swifts, hanging by each other in the
same manner as is frequently observed of bees,
which were commonly covered by the sea-wa-
ters: they appeared perfectly lifeless, but recovered on being held in his hand, or put to the
fire.

This account of the fisherman met with some
kind of confirmation by that given by Dr. Colas,
to the Royal Society, on the 12th of February,
1712-13, who, speaking of their mode of fishing
in the northern parts, by breaking holes, and
drawing their nets under the ice, said, that he
saw sixteen swallows so drawn out of the lake of
Samrodt, and about thirty out of the king's great
pond in Rosineilen; and that, at Salebitten, he
saw two swallows just come out of the water, that

could scarcely stand, from being both very wet
and very weak : with their wings hanging on the
ground, and that he had often observed the swallows to be very weak for some days after their appearance.

Dr. Owen, speaking of woodcocks and field-
fares visiting us in the winter, and then returning northward, says, " But as to cuckoos and
swallows, it is generally allowed that they sleep
in winter; having, as it is said, been found in
hollow trees and caverns; nor is this at all unlikely, though, on the other hand, I can see no
absurdity in supposing that these should go upon
a summer, as the others do upon a winter pilgri-
mage; that these pursue a lesser heat, as well as
the others fly from a greater cold.",

Willoughby, however, is of a firm opinion,
that the swallows emigrate from these climates in
the winter, and that they take their route into
Egypt and Ethiopia, in which he is confirmed by
most modern travellers; we have it, indeed, cor-
roborated in a very particular manner by a very
respectable author, in speaking of the towns of

parts of Great Britain, in the following words:
" In these towns, Southwould and Ipswich, in
particular, and so at all the towns on this coast,

place where our summer friends, the swallows,
first land when they come to visit us; and here

they may be said to begin their voyage when they go back into warmer climates. I was some years since at this place, about the beginning of October, and lodging at a house that looked into the church-yard, I observed, in the evening, an unusual number of swallows sitting on the leads of the church, and covering the tops of several houses round about. This led me to enquire, of a gentleman of the place, what could be the meaning of such a multitude of swallows having collected together, and sitting in that manner?"—' O, sir, (replied he,) you may easily perceive the reason; the wind is off the sea; for this is the season of the year, when the swallows, their food failing here, begin to leave us, and return to the country, wherever it be, from whence I suppose they came; and this being the nearest land to the opposite coast, and the wind being contrary, they are waiting for a gale, and may be said to be wind-bound.' And of the justness of this remark, I was convinced in the morning, when I found the wind had come about to the north-west in the night, and there was not a single swallow to be seen.

" To me, however, it appears certain, that swallows neither come hither merely for warm weather, nor retire merely from cold; they, like shoals of fish in the sea, pursue their prey; they are a voracious creature, and feed flying; for their food is the insects, of which, in our sum-

mer evenings, in damp and moist places, the air
is full. They come hither in the summer, because our air is fuller of fogs and damps than in
other countries, and for that reason breeds greater
quantities of insects. If the air be hot and dry,
the gnats die of themselves; and even the swallows will be found famished, and fall down dead,
when on the wing, for want of food. In the like·
manner, when the cold weather comes in, the
insects all die, and then, of necessity, the swallows quit us, and follow their food. This they do
in going off sometimes in vast flights like a cloud;
and sometimes, when the wind grows fair, they
go away a few at a time, not staying at all upon
the coast. This passing and repassing of the
swallows is observed no where so much as on this
eastern coast, namely, from above Harwich to
the east point of Norfolk, called Winterton
Ness, north, which is all right against Holland;
the passage of the sea being, as I suppose, too
broad from Flamborough Head, and the shore
of Holderness, in Yorkshire," &c.

That there have been many well-authenticated
instances of the birds being found torpid under
water, or in crevices of rocks, both here and in
some other countries, cannot be denied. But a
migration of the major part of these birds is not
to be contradicted, by what seems to be rather
the effect of chance than design. Those birds,
that have been lately hatched, and have not ac-

quired sufficient strength to accompany their companions in their Journey, may alone have been found in a torpid state.

A single instance is recorded of some swallows having with warmth and care been preserved alive through the winter, by a Mr. Pearson, of London, who, on the 14th of February, 1786, exhibited them to the society for promoting natural history. They died from neglect in the following summer.

The swallows are generally supposed to retire in the winter to Senegal, and some other parts of Africa. Dr. Russel says, that they visit the country about Aleppo, towards the end of February, where like those in England they breed. Having hatched their young, they disappear about the end of July, and returning in the beginning of October, continue somewhat more than a fortnight, and then disappear till the spring. They are found in almost all parts of the old continent, and are by no means uncommon in North America.

This bird's nest is composed of mud mixed with straw and hair, and lined with feathers. It lays four or five eggs, and has two broods in the year. The progressive method by which the young are introduced to their proper habits, is very curious. They first, but not without some difficulty, emerge from the shaft: for a day or two they are fed on the chimney-top; and then are conducted to the dead leafless bough of some neighbouring tree, where sitting in a row, they are attended by the parents with great assiduity. In a day or two after this, they are strong enough to fly, but continue still unable to take their own food ; they therefore play about near the place, where the dams are watching for flies ; and, when a mouthful is collected, at a certain signal the dam and the nestling advance, rising towards each other and meeting at an angle; the young all the while uttering such a short quick note of gratitude and complacency, that a person must have paid very little regard to the wonders of nature who has not remarked this scene.

As soon as the dam has disengaged herself from the first brood, she immediately commences her preparations for a second, which is introduced into the world about the middle or latter end of August.

Professor Kalm, in his travels into America, says, that a very reputable lady and her children related to him the following story respecting these birds, assuring him at the same time that they were all eye-witnesses to the fact:—" A couple of swallows built their nest in the stable belonging to the lady ; and the female laid eggs in the nest, and was about to brood them. Some days after, the people saw the female still sitting on the eggs; but the male, flying about the nest, and sometimes settling on a nail, was heard to utter a very plaintive note, which betrayed his uneasiness. On a nearer examination, the fe-

Interesting anecdotes.

male was found dead in the nest; and the people
flung her body away. The male then went to sit
upon the eggs; but after being about two hours
on them, and perhaps finding the business too
troublesome, he went out, and returned in the
afternoon with another female, which sat upon
the nest, and afterwards fed the young ones till
they were able to provide for themselves."

At Camerton Hall, near Bath, a pair of swallows built their nest on the upper part of the
frame of an old picture, over the chimney-piece;
entering through a broken pane in the window
of the room. They came three years successively;
and in all probability would have continued to
do so, had not the room been put in repair, which
prevented their access to it.

Another pair was known to build for two years
together, on the handles of a pair of garden
shears, that were stuck up against the boards in
an out-house; and therefore must have had their
nest spoiled whenever the implement was wanted:
and what is still more strange, a bird of the same
species built its nest on the wings and body of
an owl, that happened by accident to hang dead
and dry from the rafter of a barn, and so loose
as to be moved by every gust of wind. This owl,
with the nest on its wings, and with eggs in the
nest, was brought as a curiosity to the museum of
Sir Ashton Lever. That gentleman, struck with
the oddity of the sight, furnished the bringer with
a large shell, desiring him to fix it just where the

'owl had hung. The person did so; and the following year a pair, probably the same, built their nest in the shell, and laid eggs. The owl and the shell made a strange and grotesque appearance; and are now not the least singular specimens in that wonderful collection of the curiosities of art and nature, the Leverian Museum.

The swallow during every part of the summer is a most instructive pattern of unwearied industry and affection; for, from morning to night, while there is a family to be supported, she spends the whole time in skimming along, and exerting the most sudden-turns, and quick evolutions; avenues, and long walks under hedges, pasture fields, and mown meadows where cattle graze, are their delight, especially if there be trees interspersed, because in such parts insects most abound. When a fly is taken, a smart snap from their bill is to be heard, not unlike the noise of the shutting of a watch-case; but the motion of the mandibles is too quick for the eye.

The house-swallow is the excubitor to the house-martins and other little birds, announcing the approach of birds of prey: for as soon as a hawk or an owl appears, the swallow calls, with a shrill alarming note, all his own fellows and the martins about him; who pursue in a body, and buffet and strike their enemy till they have driven him from the place, darting down upon his back and rising in a perpendicular line in perfect security. This bird will also sound the alarm, and strike at cats, when they climb on the roofs of houses, or otherwise approach the nests.

Wonderful, observes Mr. White, is the address which this adroit bird exhibits in ascending and descending with security through the narrow passage of a chimney. When hovering over the mouth of the funnel, the vibrations of its wings acting on the confined air, occasion a rumbling like distant thunder. It is not improbable that the dam submits to the inconvenience of having her nest low down in the shaft, in order to have her brood secure from rapacious birds ; and particularly from owls, which are frequently found to fall down chimneys, probably in their attempts to get at the nestlings.

A writer in the gentleman's magazine observes, that " by the myriads of insects which every single brood of swallows destroy, in the course of a summer, these birds defend us in a great measure from the personal and domestic annoyance of flies and gnats; and what is of infinitely more consequence, they keep down the numbers of our minute enemies, which, either in the grub or winged state, would otherwise prey on the labours of the husbandman. Since then swallows are guardians of our corn, they should every where be protected by the same popular veneration which in Egypt defends the ibis, and the stork in Holland. We more frequently hear of unproductive harvests on the continent than in this country ; and it is well known that swallows are caught, and sold as food, in the markets of Spain, France, and Italy. When this practice has been very general and successful, I have little doubt that it has at times contributed to the scarcity of corn. In England, we are not driven to such resources to furnish our tables. But what apology can be made for those, and many there are, whose education and rank should have taught them more innocent amusements, but who wantonly murder swallows under the idle pretence of improving their skill in shooting game? Besides the cruelty of starving whole nests by killing the dam, they who follow this barbarous diversion would do well to reflect, that by every swallow they kill, they assist the effects of blasts, mildews,

lord of a manor should restrain his game-keeper from this execrable practice; nor should he permit any person to sport on his lands who does not refrain from it. For my part, I am not ashamed to own, that I have tempted martins to build round my house, by fixing scallop shells in places convenient for their ' pendent beds and procreant cradles'; and have been much pleased in

raises a buttress under each shell, before he ventures to form his nest on it."

We are informed by M. de Buffon that a shoemaker in Brasil put a collar on a swallow, containing an inscription to this purpose:

" Pretty swallow, tell me, whither goest thou in winter?"

3

And in the ensuing spring received, by the same courier, the following answer :

" To Antony at Athens ;—Why dost thou enquire ?"

The most probable conjecture on this story is, that the answer was written by some one who had caught the bird in Switzerland; for both Belon and Aristotle assure us, that though the swallows live half the year in Greece, yet they always pass the winter in Africa.

Mr. White informs us, that for some weeks before the swallows depart, they (without exception) forsake houses and chimneys, and roost in trees; and they usually withdraw about the beginning of October, though some few stragglers may be seen at times till the first week in November. Mr. Pennant says, that for a few days previous to their departure, they assemble in vast flocks on house-tops, churches, and trees, from whence they take their flight.

MARTIN, OR WINDOW SWALLOW,

IS inferior in size to the common swallow, and its tail less forked. The head and upper parts of the body, are of a glossy blue black : the breast and belly are white, as, is also, the rump which may be considered, as its distinctive character.

These birds begin to appear about the 16th of April: and generally for some time pay no attention to the business of nidification; but play and sport about, either to recruit from the fatigue of lse that their blood may recover its true tone and texture, after having been so long benumbed by the severities of winter. About the middle of May, if the weather be fine, they then begin to think of providing a mansion for their family.

As the martin often builds against the eave of a house, the side of a cliff over the sea, or a perpendicular wall, without any projecting ledge under, its utmost efforts are necessary to get the first foundation firmly fixed, so as to carry safely the superstructure. On this occasion the bird not only clings with its claws, but partly supports itself by strongly inclining its tail against the wall, making that a fulcrum; and thus fixed, it plasters the materials into the face of the brick or stone. But that this work may not, while soft, incline down by its own weight, the provident architect has the prudence and forbearance not to proceed too fast.; but by building only in the morning and dedicating the rest of the day to food and amusement, gives it sufficient time to dry and harden. About half an inch seems to be a sufficient layer for a day. By this method, in about ten or twelve days, a hemispherical nest is formed, with a small aperture to-

wards the top; strong, compact, and warm, and
perfectly fitted for all the purposes for which it
was intended. But nothing is more common
than for the house-sparrow, as soon as the shell
is finished, to seize on it, eject the owner, and
to line it according to its own peculiar manner.
After so much labour is bestowed in erecting a
mansion, as nature seldom works in vain, martins
will breed for several years successively in the
same nest, where it happens to be well sheltered
and secured from the injuries of the weather.
The shell or crust of the nest is a sort of rustic
work, formed of such dirt or loam as is most
readily met with, and tempered and wrought together with little pieces of broken straws to render it tough and tenacious; it is full of knobs
and protuberances on the outsides; nor is the
inside smoothed with any great exactness, but
is rendered soft and warm, and fit for incubation,
by a lining of small straws, grasses, and feathers,
and sometimes by a bed of moss interwoven with
wool.

Herein the female produces four or five
young; which, when arrived at full growth, become impatient of confinement, and sit all day
with their heads out at the orifice, where the
dams, by clinging to the nest, supply them with
food from morning to night. After this they
are fed by the parents on the wing; but this
feat is performed by so quick and almost imper-

ceptible a flight, that a person must attend very
exactly to the motions of the birds, before he is
able to perecive it.

As soon as the young are able to provide for
themselves, the dams repair their nests for a
second brood. The first flight then associate in
large flocks; and may be seen on sunny mornings and evenings, clustering and hovering
around towers and steeples, and on the roofs of

begin to take place about the first week in August. From observing the birds approaching
and playing about the eaves of buildings, many
persons have been led to suppose that more than
two old birds attend on each nest.

The martins are often very capricious in fixing
on a nesting-place, beginning many edifices
and leaving them unfinished; but (as before observed) when a nest has been once completed in
a sheltered situation, it is made to serve for several seasons. In forming their nests these in-
dustrious creatures are at their labour, in the
long days, before four o'clock in the morning:
in fixing their materials, they plaster them on
with their chins, moving the head with a quick
vibratory motion. They breed the latest of
all our swallows, never being without unfledged
young even so late as Michaelmas.

Sometimes in very hot weather they dip and
wash as they fly, but not so frequently as the

swallows. They are the least agile of all the British hirundines; their wings and tails are short, and therefore they are not capable of those surprising turns, and quick and glancing evolutions, that are so observable in the house swallows.

Their motion is placid and easy : generally in the middle region of the air ; for they seldom mount to any great height, and never sweep long together over the surface of the ground or water. They do not wander far in quest of food ; but are fond of sheltered places near some lake, or under some hanging wood or hollow vale, especially in windy weather.

As the summer declines, the flocks increase in number every day from the accession of the second broods ; till at length, round the villages on the Thames, they swarm in myriads, darkening even the face of the sky as they frequent the aits of that river, where they roost. The bulk of them retire, in vast companies, about the beginning of October ; but some have been known to remain so late as till the sixth of November : they are the latest of all the species in withdrawing. It would seem that either these are very short-lived birds, or that they undergo vast destruction in their absence, or do not return to the districts where they were bred ; for the numbers that appear in the spring, bear no proportion to those that retired in the preceding year.

Mr. Simpson, during his residence at Welton, in North America, one morning heard a noise from a couple of martins that were flying from tree to tree near his dwelling. They made several attempts to get into a box or cage fixed against the house, which they had before occupied; but they always appeared to fly from it again with the utmost dread; at the same time repeating those loud cries which first drew his attention. Curiosity led this gentleman to watch their motions. After some time a small wren came from the box, and perched on a tree near it; when her shrill notes appeared to amaze her antagonists, Having remained a short time, she flew away. The martins took this opportunity of returning to the cage; but their stay was short. Their diminutive adversary returned, and made them retire with the greatest precipitation. They continued manœuvering in this way the whole day; but the following morning, on the wren's quitting the cage, the martins immediately returned, took possession of their mansion, broke up their own nest, went to work afresh with extreme industry and ingenuity, and soon barricaded their doors. The wren returned, but could not now re-enter. She made attempts to storm the nest, but did not succeed. The martins, abstaining from food nearly two days, persevered during

and the wren, finding she could not force the

works, raised the siege, quitted her intentions, and left the martins in quiet possession of their nest.

SAND-MARTIN, OR SAND-SWALLOW,

WHICH is less than the preceding species, is about four inches and three quarters in length. The upper parts of the body are mouse-colored, and the breast and belly white, with a mouse-colored ring as a collar.

This bird is common about the banks of rivers and sand-pits, where it digs itself a round and regular hole in the sand or earth : this is hori-zontal, serpentine, and generally about two feet deep. At the farther end of this burrow, the bird constructs her rude nest of grass and fea-thers. " Though one would at first be disin-clined to believe (says Mr. White) that this weak bird, with her soft tender bill and claws, should ever be able to bore the stubborn sand-bank without entirely disabling herself; yet with these feeble instruments have I seen a pair of them make great dispatch; and could remark how much they had scooped in a day, by the fresh sand which ran down the bank, and which was of a different colour from what lay loose and had been bleached in the sun. In what space of time these little artists are able to mine and finish

but it would be a matter worthy of observation,

make such remarks. This I have often taken
notice of, that several holes of different depths
are left unfinished at the end of the summer. To
imagine that these beginnings were intentionally
made in order to be in the greater forwardness
for next spring, is allowing, perhaps, too much
foresight to a simple bird. May not the cause
of these being left unfinished arise from the birds
meeting in those places with strata too harsh,
hard, and solid, for their purpose; which they
relinquish, and go to a fresh spot that works
more freely? Or may they not in other places
fall in with a soil as much too loose and moulder-
ing, liable to flounder, and threatening to over-
whelm them and their labours? One thing is
remarkable: that, after some years, the old
holes are forsaken, and new ones bored; perhaps
because the former habitations were become foul
and fetid from long use, or because they so
abounded with fleas as to become untenable."

The sand-martin appears in this country about
the same time as the swallow, and lays from four
to six white and transparent eggs. These birds
seem not to be of a very sociable disposition,
never (at least in England) congregating in the
autumn. They have a peculiar manner of flying: flitting about with odd jerks and vacillations not unlike the motions of a butterfly.

These birds are, as already intimated in the above extract from Mr. White, so strangely annoyed with fleas, that these vermin have been sometimes seen swarming at the mouths of their holes, like bees on the stools of their hives.

THE SWIFT, OR BLACK MARTIN,

IS the largest of the kind known in these climates, being often near eight inches long, with an extent of wing near eighteen inches, though the entire weight of the bird is not more than an ounce. The whole plumage is a sooty black, except the throat, which is white. The feet, which are so small, that the actions of walking and rising from the ground seem very difficult, are of a particular structure, all the toes standing forward. The least toes consist of only one bone: the others of two each; in which they differ from the toes of all other birds: this is, however, a construction nicely adapted to the purposes in which the feet of these birds are employed. Nature has also made it ample compensation for its difficulty in walking and rising, by furnishing it with abundant means for an easy and continual flight. It spends more of its time on the wing than any other swallow, and its flight is more rapid. It breeds under the eaves of houses, in steeples, and other lofty buildings; and makes its nest of grass and feathers.

. This bird visits us the latest, and leaves us the

arrive before the beginning of May, and seldom remains later than the middle of August. It is the most active of all birds : being on the wing,

in the day ; withdrawing to rest, in the longest days, about a quarter before nine in the evening, some time after all the other day-birds are gone. Just before they retire, large groups of them assemble high in the air, screaming, and shooting about with wonderful rapidity. This bird is, however, never so alert as in sultry louring weather ; when it expresses great alacrity, and calls forth all its powers.

The swifts, in hot mornings, collect together in little parties, and dash round the steeples and churches, squeaking at the same time in a very clamorous manner. These are supposed to be the males, serenading the sitting hens; as they seldom make this noise till they come close to the walls or eaves, and those within always utter a faint note of complacency. When the hen has been occupied all day in sitting, she rushes forth, just before it is dark, to relieve her weary limbs ; she snatches a scanty meal for a few minutes, and then returns to the task of incubation.

These birds differ from all the other British hirundines, in breeding but once in the summer, and in producing no more than two young ones at a time.

These birds, when shot while they have young, are found to have a little cluster of insects in their mouths, which they pouch and hold under their tongue. In general, they fly and feed higher in the air than the other species. They also range to vast distances; for motion is but a slight labour to them, endowed as they are with such wonderful powers of wing. Sometimes in the summer, they may, however, be observed hawking very low, for hours together, over pools and streams, in search of the cadew-flies, may-flies, and dragon-flies, that frequent the banks and surface of waters, and which afford them a plentiful and succulent nourishment. Sometimes they pursue and strike at birds of prey when they are sailing about in the air; but they do not express so much vehemence and fury on these occasions as the swallows.

The voice of the swift is a harsh scream; yet there are few ears to which it is not pleasing, from an agreeable association of ideas, since it is never heard but in the most lovely summer weather. These birds never settle on the ground unless by accident, from the difficulty they have in walking, or rather (as it may be called) in crawling; but they have a strong grasp with their feet, by which they readily cling to walls and other places they frequent. Their bodies being flat, they can enter a very narrow crevice; and where they cannot pass on their bellies, they will turn up edgewise to push themselves through.

The main body of these birds retire from this country before the middle of August, generally by the tenth, (which is but a short time after the flight of their young,) and not a single straggler is to be seen on the twentieth. This early retreat is totally unaccountable, as that time is often the most delightful in the year. But what is yet more extraordinary, they begin to retire still earlier in the most southerly parts of Andalusia; where they can by no means be influenced by any defect of heat, or even (as one would suppose) of food. This is one of those incidents in natural history, which not only baffle our researches, but elude all our conjectures.

A pair of swifts were, in the month of February 1766, found adhering by their claws, and in a torpid state under the roof of Longnor Chapel, in Shropshire: on being brought to the fire, they revived, and moved about the room.

THE ESCULENT SWALLOW.

THIS is said to be less in size than the wren. The bill is thick. The upper parts of the body are brown, and the under parts whitish. The tail is forked; and each feather is tipped with white. The legs are brown.

This bird's nest is exceedingly curious; and is composed of such materials that it is not only edible, but is accounted among the greatest dainties by the Asiatic epicures. It generally weighs about half an ounce ; and is in shape like a half-moon, or, as some say, like a saucer, with one side flatted, which adheres to the rock. The texture somewhat resembles isinglass, or fine gum-dragon : and the several layers of the component matter are very apparent ; it being fabricated from repeated parcels of a soft slimy substance, in the same manner as the martins form their nests of mud. Authors differ much as to the materials of which this nest is composed : some suppose it to consist of sea-worms of the Mullusea class ; others, of the sea-qualm (a kind of cuttle fish), or a glutinous sea-plant called Agal-agal. It has also been supposed that the swallows rob other birds of their eggs ; and after breaking the shells, apply the white of them in the composition of these structures.

The best sort of nests, which are perfectly free from dirt, are dissolved in broth, in order to thicken it ; and are said to give it an exquisite flavour. Or they are soaked in water to soften them ; then pulled to pieces ; and, after being, mixed with ginseng, are put into the body of a fowl. The whole is then stewed in a pot, with a sufficient quantity of water, and left on the coals all night. On the following morning it is ready to be eaten.

These nests are found in vast numbers in certain caverns of various islands in the Soolo Archipelago. The best kind sell in China at from one thousand to fifteen hundred dollars the picle, (about 25 pounds) ; the black and dirty ones for only twenty dollars. It is said that the Dutch alone export from Batavia one thousand pieles of these nests every year ; which are brought from the islands of Cochin-China, and those lying east of them. Notwithstanding the many luxuries imported by us from the East, these nests are yet so scarce in England, that they have not found their way to our tables.

Sir George Staunton, in his " Account of the Embassy to China," gives the following description of this bird's nest :—" In the Cass (a small island near Sumatra) were found two caverns, running horizontally into the side of the rock ; and in these were a number of those birds'-nests so much prized by the Chinese epicures. They seem to be composed of fine filaments ; cemented together by a transparent viscous matter, not unlike what is left, by the foam of the sea, upon stones alternately covered by the tide, or those gelatinous animal substances found floating on every coast. The nests adhere to each other, and to the sides of the cavern ; mostly in rows, without any break or interruption. The birds that build these nests are small grey swallows, with bellies of a dirty white. They were flying about in considerable numbers ; but were so small, and their flight was so quick, that they escaped the shot fired at them. The same sort of nests are said to be also found in deep caverns

Sir George Staunton's account.

at the foot of the highest mountains in the middle of Java, at a distance from the sea: from which source it is thought that the birds derive no materials, either for their food, or the construction of their nests; as it does not appear probable they should fly, in search of either, over the intermediate mountains, which are very high, or against the boisterous winds prevailing thereabout. They feed on insects, which they find hovering over stagnated pools between the mountains, and for the catching of which their wide-opening beaks are particularly adapted. They prepare their nests from the best remnants of their food. Their greatest enemy is the kite; who often intercepts them in their passage to and from the caverns, which are generally surrounded with rocks of grey limestone, or white marble. The nests are placed in horizontal rows, at different depths, from fifty to five hundred feet. The colour and value of the nests depend on the quantity and quality of the insects caught; and, perhaps, also on the situation where they are built. Their value is chiefly ascertained by the uniform fineness and delicacy of their texture; those that are white and transparent being most esteemed, and fetching often in China their weight in silver.

"These nests are a considerable object of traffic among the Javanese, many of whom are employed in it from their infancy. The birds, after having spent nearly two months in preparing

their nests, lay each two eggs, which are hatched in about fifteen days. When the young birds
become fledged, it is thought the proper time to
seize upon their nests; which is done regularly
three times a year, and is effected by means of
ladders of bamboo and reeds, by which the people descend into the caverns: but when these
are very deep, rope ladders are preferred. This
operation is attended with much danger, and
several perish in the attempt. The inhabitants
of the mountains generally employed in this business, begin always by sacrificing a buffalo;
which custom is observed by the Javanese on the
eve of every extraordinary enterprize. They
also pronounce some prayers, anoint themselves
with sweet-scented oil, and smoke the entrance
of the cavern with gum-benjamin. Near some
of the caverns a tutelar goddess is worshipped;
whose priest burns incense, and lays his protect-
ing hands on every person preparing to descend.
A flambeau is carefully prepared at the same
time, with a gum which exudes from a tree
growing in the vicinity, and which is not easily
extinguished by fixed air or subterraneous va-
pours."

Description.

CHAP. III.

"These, far dispersed, on tim'rous pinions fly
To sport and flutter in a kinder sky."

GOLDSMITH.

THE CROWN·BIRD.

THERE being several of the feathered tribes in different parts of the world with whose figures or manners, naturalists confess themselves almost unacquainted, we shall devote this chapter to a few particulars of them.

The crown-bird (with which we shall first begin) is a fine stately East India fowl, about the size of an English turkey; the body is covered with long slender feathers resembling hair, of a dark green colour, with a purplish cast on the sides and back, with a few broad stripes of red upon the wings, tending downwards; the thighs are a sort of buff colour, the claws black. It has a large bluish, or gold-coloured tuft on the top of the head, which grows up in shafts or stalks, with little balls upon the tops, that bear a trifling resem-

blance to an earl's coronet, according to some;
others say it is more like the tuft on the head of
a Virginian nightingale. A little above the bill
upon the forepart of the head, is a small red
comb, and two red marks, resembling ears, on
each side of the head; the bill is short and thick,
bending downward, of a yellow colour.

This seems to be the bird described by **Mr.**
Tavernier in his travels into India; great numbers of which are found in the territories of Cam-
baya, Broudra, &c. which in the day-time walk
about the fields, but in the night roost upon the
trees. The flesh of the young ones, he says, is
white and well tasted. In those parts where the
Mahometans govern, they may be caught without
danger; but in those territories where idolatrous
rajahs are masters, it is very dangerous to kill
them, or any other bird or animal; for the ba-
nians count it sacrilege, and will severely punish
any they can seize. They whipped a Persian
merchant to death, and took all his money, to the
value of three hundred thousand rupees, for
shooting a peacock. ·

The Mexican crown bird has a thick short bill,
of a sort of flesh colour, or tawny, with a large
crest of green feathers upon the head, which it
raises and falls at pleasure. The head, neck,
back, breast, and part of the belly and thighs,
are of a brownish dusky colour. The four first
quill-feathers of the wings are a fine scarlet, the

last having fine long white marks upon the outward web: the rest of the quill-feathers and the
tail are purple, as are also the covert and scapular
feathers of the wings, with a fine mixture of
green, interspersed through the whole. The
legs and feet are bluish, or lead colour. It is in
size pretty near to that of a fieldfare.

Barbot, in his " Description of his voyage to
South Guinea," describes this as a fine bird, of various colours, as white, black, brown, red, sky colour, blue, &c. having a long tail, the feathers
whereof, he says, the blacks wear on their heads.
He describes some of them which are of a gold
colour, and others with charming blue tufts on
their heads, much in the form of a Virginian
nightingale's.

THE CARASOW,

SO called from a part of the West Indies, from
whence it is brought; the Indians give it the
name of the mountain bird, and some travellers
call it a wild turkey; it is a bird which is easily
made tame and sociable, so as to accompany with
other fowls. This beautiful bird is black upon
the head and neck, resembling velvet, and has a
high crest of curious ruffled black feathers like a
half circle, which rises spirally from the top of its
head, with a white circle running across them;
these it can erect, or let fall at pleasure. The rest

of the body, excepting the lower part of it, of the cock, are black; that of the hen rather of a dusky brown; the tail is black, with four bars of white running across it near the extremity, at equal distances. The bill is thick on the upper mandible, of which there is a round excrescence as big as a hazle nut; the eyes are black, and the legs pretty long, and the size of its body not a great deal less than a common turkey.

M. W. in his " Description of the kingdom of Mosqueto", calls it a small Indian turkey, and says they are very welcome game to the hungry traveller, who may shoot all he meets with, one after another, they being so very tame, that they will scarcely fly away : they keep frequently ten or a dozen in a flock, and are excellent meat.

The author of the " Buccaniers of America" takes notice of this bird under the name of Oecos, and says it very much resembles the European turkey, and that the feathers of some of the male kind are inclining to red, but those of the female to black.

Dr. Gemelli says, they have abundance of them in

killed; " for when one falls, (he says) there is no danger of the rest flying away at the noise of the gun."

THE MANAKIN.

THIS is a tribe of small but very pretty birds, the largest of which are inferior in size to the sparrow, and the others are as small as the wren. The general characters which are common to all the varieties, are a short, straight bill, compressed on the sides towards the end; the upper mandible convex above, and somewhat serrated on the edges, rather longer than the lower, which is flat and perfectly straight. All these birds, likewise, have a short tail of a square figure.

The natural habits of these birds have not yet been observed with sufficient accuracy to enable us to give an exact detail of them. The only naturalist that has written on this subject is M. Sonnini, who, during his residence in South America, saw a great number of manakins in a state of nature. They inhabit the extensive forests in the hot climates of that continent, and this residence they never exchange for the open plain, or the neighbourhood of human habitations. Their flight, though rapid, is always short and low; they never perch on the summits of trees, but on branches of a moderate height, feeding upon the smaller kinds of wild fruits, and on insects.

They are usually seen in companies of eight or ten of the same species; sometimes they are mingled with troops of other species, and even

with companies of small birds of a different kind.
They are observed to assemble in this manner in
the morning, which seems to inspire them with
joy, to judge by their agreeable warbling at
such times. This expression of delight is evidently produced by the refreshing coolness of
the morning; for they are silent during the
day, and endeavour to screen themselves from
the noon tide heat by separating, and retiring
singly into the most shady parts of the forests,
where they remain till the following morning.
They, in general, prefer moist and cool situations to such as are dry and hot; yet, they never
frequent either marshes or the banks of waters.

The name of manakin was given to these birds
by the Dutch colonists of Surinam. Naturalists
have discovered six different species of this tribe.
That which is the largest, and is called the crested
manakin, or, by Edwards, the blue-blacked manakin, is about four inches and a half long, and is nearly as large as a sparrow. The upper part of the head
is covered with feathers of a beautiful red colour;
these are longer than the others; the bird can erect
them at pleasure, and they produce the appearance of a crest; the back and the upper wing-coverts are of a charming blue; the rest of the plumage is of a velvet black. The feet are red, the
bill black, and the iris of the eye of a beautiful
sapphire.

THE-RED LEGGED HORSEMAN.

THIS singular and uncommon bird, is supposed to be a native of the low countries; very few of them have been seen in England, and those few in the eastern parts of Essex; where they acquired their name, which cannot, however, be justly applied, on account of the colour of their legs, for although those of the cock are of a pale red, those of the hen are green. The top of the head and the neck are of a fine light brown, the bill is slender, and near two inches long, of a reddish colour at the base, and black at the point. The covert-feathers in the middle of the wings black, but downwards, within two inches of the end of the tail, they are of a brown colour edged, with white. The legs are pretty long, the claws small, and black. They are said to be about sixteen or eighteen inches from the point of the bill to the end of the claws; and in breadth, when the wings are fully expanded, two feet. Its weight about half a pound.

THE BRASILIAN NIGHT BIRD.

SOME say that this bird is seldom larger than a fowl, while others maintain it is bigger than a goose; the probability is, that there are some not

S

only of both, but also of the intermediate sizes.
It is an ill-favoured bird; the head much resembles that of a cat: it has a crooked bill, the upper
mandible hanging a good way over the under,
the eyes are large and sprightly, shining like crystal, the inner circle appears of a whitish yellow;
it has two large tufts of feathers near two fingers
long, upon the head, resembling ears; the tail is
so short, that it is not seen when the wings,
which are very long, are closed.

The thighs and legs are covered with short
down, or feathers, as low as the feet, which are
armed with strong crooked claws, near three
inches long, and very sharp. The feathers of
the whole body are of a gold colour sprinkled up
and down with black and white spots.

THE BEE-EATER,

WHICH very much resembles the kingfisher,
and is about the size of a common black-bird, has
a large oblong head, the feathers at the base of the
upper chap, are white shaded with a sort of green
amongst it; from the corners of the bill, along each
side of the head, there is a black stroke extending itself beyond the eyes; contiguous to which,
on the under part of the head, the feathers are of
a pale yellow; the belly, neck, and breast are of
a bluish green; the scapular feathers in some
are blue; in others green, with a mixture of red.

The large wing-feathers are of a sort of orange
colour, with black tips, intermixed with some
few green ones; the rest of the wing-feathers are
in some birds more red, in others more blue. The
tail is upwards of three inches long, and consists of
about twelve feathers; the two middlemost of
which are considerably longer than the rest, and
end in sharp points; the colour of the tail varies;
in some it is green, in others, blue; the under side
of a dark brown. The bill is large and near two
inches, but bends downwards, and is a good deal
more arched than the kingfisher's: the tongue is
slender but appears rough towards the end and jagged as if it were torn. The eyes are in some hazle; in others of a fine red colour. The legs and
feet very nearly resemble those of the king-fisher,
the toes being joined much in the same manner;
they are generally of a blackish colour, sometimes of a sort of brown or dusky red; the claws
black.

Bellonius says it is common in the island of
Crete, and sometimes is seen in some parts of
Italy, but is unknown upon the continent of
Greece. It feeds not only on bees, but upon ci-
cada, beetles, grasshoppers, and other insects, and
also on the seed of nipplewort, bastard parsley,
turnips, &c. They fly in flocks, and frequent the
mountains that bear wild thyme.

Aristotle says they build in hollow places or caverns three or four cubits deep, and lay six or
seven eggs.

Bengal bee-eater.

The Bengal bee-eater is about the size of the former; and has a black bill, thick at the base, bending downwards, near two inches long, the eyes a fine red; a black stroke is extended on each side of the head, which begins at the corners, of the mouth, and passes beyond the eyes. The base of the upper chap, and under the chin, is covered with bright pale blue feathers; the upper and back part of the head, of a dusky yellow; the back and wings of the same colour, only shaded pretty strongly with a green; the tips of the quill-feathers, brown, the breast and belly green; the thighs and under part near the vent, of a pale yellow, with a small green mixture. The tail consists of the same number of feathers as the former; the outermost on each side, are, of a green and yellow mixture, about three inches in length; the two middlemost twice that, length, ending in sharp points, of a brown or dusky colour; the legs and feet black, much like the former.

THE WRY-NECK,

WHICH is nearly the size of a common lark, has a short lead-coloured bill, something less triangular than the rest of the birds of this kind, and has a round tongue, which ends in a sharp and bony substance, and pointed like a thorn, with which it generally strikes the ants that are its food, which, by the gluttinous matter it contains, and the bird contracting its tongue, it swallows them without ever touching them with its bill.

Its plumage, in general, is very elegant and curiously coloured; the upper part of the body being variegated with a great many colours, and a beautiful sharp line that runs from the crown of the head along the middle of the back. As a means of distinguishing the cock from the hen, Mr. Denham has been so curious as to observe that the black line upon the cock runs no farther than the upper part of the neck, but that of the hen, he says, reaches almost to the very bill; also, that the cock's belly is naked, as is the hen's when she sits; from whence he concludes, it takes its turn in incubation. The lower part of the belly and the throat are yellowish, with some black transverse lines : the prime feathers of the wings are spotted with large white spots; the rump of an ash colour. The tail is about two inches long, crossed at distant intervals with black, and sprinkled with little dark coloured specks, but does not bend inwards as other woodpeckers do. The legs and feet are short, and the claws disposed in the same manner as other birds of this kind.

There is something ridiculously odd in his gesture, it frequently turning its head quite back to its shoulders; from whence the ancients have given it the name of torquilla. The body of the hen is of a paler, or more cinerous colour than the cock.

THE HORNBILL.

IT is said there are eleven species of the horn-bill; it is nearly allied to the toucan, and indeed seems to hold the same place in the warm climates of the old continent, as the toucan does in the new. The distinguishing characteristic of this genus is an immense bending bill, with frequently a large protuberance in the upper part of it, resembling an additional bill.

The rhinoceros hornbill, or rhinoceros bird, is nearly as large as a turkey; the bill is ten inches long, and two and a half thick at the base. On the upper part is an appendage as large as the bill itself, and turning upwards, which measures eight inches in height. There is nothing else remarkable in the bird, as the general colour of the plumage is black. This bird is found in most parts of the East Indies, where, like the raven, it feeds upon carrion.

The helmit hornbill is remarkable for having the same prominence of a conical form; and in the Philippine isles there is a species, the horn of which reaches backwards beyond the eyes, ending in two angular points, which produce the effect of a bird with two horns.

The pied hornbill of Malabar is distinguished ;
from the rest of its kind by the breast, belly, and
part of the wings, being white, the remainder
of the body is, like the rest of these animals,
black.

THE WAXEN AND CARUNCULATED CHATTERERS.

THE waxen or Bohemian chatterer is the
size of a large lark, eight inches long. Its head is
adorned with a beautiful pointed crest. The upper parts of the body are of a reddish ash colour:
the breast and belly of a pale purplish chesnut ;
a black streak passes over each eye ; the chin
also and quills are black. Their native country
is Bohemia, whence they wander in flocks all
over Europe, and were formerly superstitiously
considered as a presage of a pestilence. They are
seldom seen in the south parts of Britain.

The carunculated chatterer is a native of
Cayenne and Brasil. It is about twelve inches
long. The plumage of the male is of a pure
white, except a tinge of yellow on the rump,
quills, and tail. The female has the upper parts
of the plumage olive grey, and the lower parts
grey, edged with olive. Both have a fleshy ca-
runcule at the base of the bill, which projects over
it like that of a turkey-cock. Their voice, as is
that of all the kind, is so loud and noisy, that

t ey may be heard at the distance of half a league.

There are about ten species of the chatterers, but very little or nothing of their habits is known.

THE GRACKLE.

THERE are about eleven species of the grackle inhabiting America and the tropical climates, some of them the size of a magpie, others about that of a blackbird. Their general plumage is black. They live on maize, fruits, and insects; but one species in the Philippine Islands, which is called from its beauty, the Paradise grackle, is remarkable for its being an extraordinary destroyer of grasshoppers. It stands upon record, that the inhabitants of the Isle of Bourbon, being greatly infested with that insect, imported a pair of these birds, which presently relieved them from that pest. In process of time, however, the grackles became very numerous, and the inhabitants thinking them injurious, proscribed them by an edict, on which the grasshoppers increased so fast upon them, that they were obliged to send for more, which soon dispatched every grasshopper on the island.

The boat-tailed grackle is a native of Jamaica. Its plumage is black, and it is remarkable for the feathers of its tail, forming a hollow like a boat

on the upper surface, so that it may be compared.
to a hen's tail with the underside turned uppermost. This bird is the size of a cuckoo.

THE DRONTE, OR DODO.

THE body of this unwieldy-and inactive bird
is almost round, and covered with grey feathers;
it is just barely supported upon two short thick
legs, while its head forms a figure so extraordinary, that it might be taken for the grotesque
production of the painter's imagination. Supported by a neck, thick and pursy, this head
consists almost entirely of an enormous beak,
the two mandibles of which open far beyond the
eyes, that are large, black, prominent, and surrounded with a white circle. The bill, therefore,
is of extraordinary length, not flat and broad,
but thick, sharp at the end, and each chap
crooked in opposite directions so as to resemble
two pointed spoons laid together. In short, the
appearance of this bird denotes the most extreme
stupidity and voracity, and its deformity is still
more increased by a border of feathers round the
root of the beak, which terminates in a point on
the forehead, and envelopes the face so as to give
the appearance of a hood or cowl.

The feathers of this bird are in general very
soft; grey is their predominant colour, but darker

on all the upper part of the body and the legs, and lighter on the breast, belly, and the whole under part of the body. The feathers of the wings are variegated with yellow and white, as are likewise those of the tail, which appear to be curled, and are very few in number. Clusius reckons only four or five. The feet and toes are yellow, and the nails black. Each foot has four toes, three before and one behind, and the claw of the latter is considerably the longest.

Bulk, which in the generality of animals implies strength, in the dronte contributes only to inactivity. The ostrich and the cassowary are no more able to fly than the dronte, but they supply that defect by their speed in running. This bird, on the contrary, appears to be oppressed by its own weight, and scarcely able to urge itself forward. It seems among birds what the sloth is among quadrupeds, an unresisting animal, equally incapable of flight or defence. It is furnished with wings covered with soft ash-coloured feathers, but they are too short and too weak to assist it in flying: it has a tail, but this tail is disproportioned and misplaced.

This bird is a native of the isle of France; and the Dutch, by whom it was first discovered there, called it in their language the *nauseous bird*, both on account of its disgusting figure and the bad taste of its flesh. Succeeding observers, however, contradict this report, and assert that its flesh is good and wholesome eating.

Two other birds have been mentioned by travellers, the one under the name of the solitary, and the other the bird of Nazareth; but there is great reason to believe they are merely varieties of the species of the drónte. The first is found in the Isle of Roderique, and its description scarcely varies from the one we have just treated of; but from being more known some particulars are mentioned which have not been so clearly ascertained of the dronte; such, for example, that they lay but one egg, and upon which they sit seven weeks; that they hatch towards the end of the year, and are much sought after by the inhabitants, from March to September, when they are very fat, and excellent food.

The nazarine, or bird of Nazareth, is found in the Isle of France, though supposed to be a native, and which indeed the appellation given it seems to imply, of the Isle of Nazareth. It is described as being larger than a swan, with the bill bent a little downwards; that the body of it is covered over with a blackish down, but that it has some frizzled feathers on the wings and rump.

THE TANGARAS, INCLUDING THE ORGANIST.

IN the sultry regions of America is found a very numerous tribe of small birds, some of

which, in Brasil, are denominated tangaras;
and naturalists have applied the name to all the
varieties composing that family. Most travellers
have considered these birds as a species of sparrows, and in fact they differ from the European
sparrows only in colour, and in having the upper
mandible of the bill serrated on each side towards
the end. In other respects they resemble the
sparrow in their conformation, and their habits
and manners are almost the same. Like the latter their flight is short and low; the voice in
most of the species is disagreeable; and they
must likewise be classed among the granivorous
birds, since they live only on very small fruits.
They are as familiar as the sparrow, residing
principally near the habitation of man, and asso-
ciating together in numerous flocks. They are
fond of dry, open situations, and are never found
in marshy countries. These birds lay two, and
very seldom three eggs.

The whole family of tangaras, of which there
are about thirty different species, exclusive of
varieties, appears to be confined to the New
World. To this class of birds belongs the organist, which is a native of the Spanish part of
St. Domingo, and has received its name from
sounding all the notes of an ascending octave.
The organist is so extremely shy, and so expert
at concealing himself, that it is extremely difficult to discover or to kill him. He turns round
a branch according as the fowler changes his

place, with such dexterity, that it is impossible to perceive him. Thus, though frequently there may be several of these birds on a tree, yet he cannot discern a single one, so attentive are they to keep themselves concealed.

The length of the organist is about four inches, the plumage of the head and neck is blue; the back, wings, and tail, are black, changing to a dark blue; the breast, rump, and all the under parts of the body, are of an orange yellow.

In the history of Louisiana, by M. le Page Dupratz, is the description of a small bird which he calls the bishop, and which we conceive to be the same with the organist. "The bishop," says he, " is rather smaller than a canary, and its plumage is blue, approaching to a violet colour, from which circumstance it derives its name. It feeds upon several sorts of grain, particularly a kind of millet that grows spontaneously in the country. Its throat is so soft and flexible, and its note so tender, that those who have once heard it, become much more sparing of their praises of the nightingale. Its song continues during the space of a *Miserere,* all which time it does not appear to take breath. It then rests twice as long, and instantly begins again; and this alternate singing and resting lasts for two hours."

Description.

THE SPOTTED AND FANTAILED FLYCATCHERS.

FLYCATCHERS are with us summer birds only, and take their name from feeding upon insects. The spotted flycatcher, however, eats

the cherry-sucker. It is, in general, of a mouse-colour, the head spotted with black, and the wings and tail edged with white. The pied fly-catcher is less than a hedge-sparrow, and is known by a white spot on the forehead.

The fantailed flycatcher is a native of New Zealand. It is about the size of the bearded tit-mouse, may be easily tamed, and will sit on any person's shoulder to pick off the flies. The whole head is black, with a white collar; the upper parts of the body olive brown; the under parts yellowish nut colour, and the tail white, except the two middle feathers, which are black.

THE OTIS, OR TARDA,

IS of the size of a large cock, has only three claws on a foot, an oblong head, full eyes, a sharp bill, a bony tongue, and a slender neck. Bellonius and Gessner describe it as much larger and stronger and as weighing sometimes thir-

Seemingly a sort of bustard.

teen pounds and a half. The head, which is but indifferently shaped, is of an ash-colour, as is likewise the neck down to the breast. It has a strong bill, and a serated or a saw-like tongue, sharp on both sides, and hard towards the end; with so wide an ear, or auditory duct, that the end of a finger may be introduced. And upon examining under the feathers, two cavities will appear, one towards the bill, and the other leading directly to the brain. It has a plump round breast, is covered with white feathers on the belly, and to the middle of the thighs, and has a great many dark brown and blackish spots on the back. The larger feathers in the wings are white, but black towards the end, and at the roots red. Those of the tail are of a dark red, and adorned with a great many fine streaks and black spots on the outside, and on the inside with red. The legs are a foot in length, pretty thick and scaly.

This seems to be a sort of bustard, such as are found upon the hills, and in the woods in the northern parts of Germany. The flesh is said to resemble that of the pheasant, and was so acceptable to the Emperor Caligula, that, as Suetonius relates, he would have had it offered in sacrifice in his temple.

THE BENGAL QUAIL.

THIS beautiful bird, which is larger than the European quail, has a dark brown bill; the top part of the head is covered with black, like a cap, under which there runs a large yellow streak,

the back part of the head; the eye is encompassed with a large black line, which reaches from the corner of the mouth to the other side of the head, under which there runs a white streak, or line. The under parts of the body are of a yellowish or buff colour, except that part next the tail, which is spotted with red. The hinder part of the neck and back, with the covert feathers of the wings, are of a yellowish green, except a large division of a pale bluish green, upon the pinion of the wings, and another pretty much the same upon the rump; the legs and feet are a sort of orange colour, the claws of a dark red.

Beauplan, in his description of Ukraine, in Tartary, says there is a sort of quails in those parts, with blue feet, which are present death to any that eat of them.

M. Misson, in his voyage to Italy, observes, that vast quantities come into those parts every spring from the African shore; and that they are o tired with their long voyage, that they will settle upon the ships they first light on, from whence they are taken with very little trouble. It is surprising how a bird which has no very strong wing, should be able to continue so long a flight. Josephus says the Arabian Gulph breeds more quails than any other place. Varro and others remark, that such large numbers have in the spring time lighted upon ships at sea, in their passage from one climate to another, as to sink the ships; and that an hundred thousand of quails and swallows together have been taken in a day. And Diodorus Siculus gives much the same account of taking them at Rhinoculara, on the edge of the wilderness, where the children of Israel were fed.

THE LAND RAIL

IS from the point of each wing, when extended, about nineteen or twenty inches, and weighs about six or seven ounces. The bill is about an inch long, and very much resembles that of a water-hen; the under mandible is of a dusky colour, the upper more whitish. The body of this bird is narrow, and seems compressed on each side; the chin, breast, and lower part of the belly are white, upon the head are two broad black lines; it has also a white line that passes from the shoulders resembling that of a moorhen ; the throat is of a sort of brown or dirty

are black, but the sides are rather reddish, or ash-coloured ; it has transverse white lines, running across the thighs. Some of the wing-feathers, especially the lesser rows, are of a deep yellow ; the tail is about two inches long, the legs are bare of feathers above the knees, the feet of a whitish colour. It has a stalking gait, and is by the Italians called the great quail, or the king of quails, and is said to be the leader or guide of those birds, from one place to another. They are said to feed upon snails, worms, and all kinds of small insects.

They are but rarely seen in England, but are very common in many parts of Ireland.

THE ATTAGEN.

THIS bird has a short black bill, hooked, and sharp at the end. The body is of various colours. The head is quite beautiful, and adorned with a fine tuft, or top-knot, of a brownish colour, chequered with black and white spots. It has black eyes, with a brown circle, and the skin of the eye-lid is scarlet. The throat, or the part underneath the bill, is covered with some very fine longish feathers, which hang down in the manner of a beard. It has a long neck, which, like the rest of the body, is slender and taper, of an ash-colour, and diversified with white and black spots. The foremost claws of the feet are pretty long, but the hindermost are shorter, and all of them are provided with sharp crooked talons.

What was the true attagen of the ancients is not very well agreed among authors. Alexander Myndius describes it to be a little larger than a partridge, to be full of spots of different colours down the back; of a reddish brown with short wings, and a plump heavy body. But Gessner takes it to have been the gallina corylorum, or mountain partridge. Bellonius thinks it was of the quail kind; and Julius Alexandrinus relates, that he saw one that was brought from Spain, that had a longer neck and legs, and was not spotted in the same manner. One was seen at Florence, with a black bill, but reddish at the end; black eyes, with an ash-coloured circle; spotted with white on the belly; of a reddish brown colour on the back, chequered with black spots; and with dark brown feet. Aldrovandus says, that a bird was brought from the mountains in Sicily, and affirmed to be the true attagen, which in bigness, and almost every other particular, exactly resembled a pheasant; their flesh is of a delicious taste.

These birds are reported to purge themselves with henbane; for which reason, it is said, none are to be seen in countries which want that plant. They are found in some parts of Crete, and in Cyprus they are bred tame. But none can com-

pare in goodness with those of Rhodes and Ionia.
They are likewise to be met with in the Pyre-
nean mountains on the side of Spain, in Au-
vergne, in France, and on the Alps.

They feed on all sorts of grain and fruits; and
will call over their own name, as well as their
voice will permit, and sing. But as Pliny and
Ælian inform us, they lose their voice when
taken; and what is equally remarkable, recover
it again as soon as set at liberty.

CHAP. IV.

Amid the cool translucent rill
  That trickles down the glade,
They bathe their plumes, they drink their fill,
  And revel in the shade.

GREAVES.

AQUATIC BIRDS.

HAVING treated on the several land birds of this and other countries, we now proceed to consider another description of the feathered race.

There are three divisions of water birds. Some are appointed by nature to reside on the land; others are destined to sail on the water, and to an intermediate tribe, (a sort of amphibious animals), the confines of the two elements have been allotted.

The general conformation of aquatic birds, exhibits fully the fitness of their destination to that element in or near which their lives are entirely spent. The body of the swimmers is arched beneath, and bulged like the hulk of a ship; and

this figure was perhaps copied in the first construction of vessels; their neck, which rises on a
projecting breast, represents the prow; their
short tail, collected into a single bunch, serves as
a rudder; their broad and palmated feet perform
the office of oars; and their thick down glistening
with oil (which entirely invests them) is impene-
trable by humidity, and at the same time enables
them to float more lightly on the surface of the
water. The habits and economy of these birds
correspond also with their organization: they
never seem happy but in their appropriate element; they are averse to alight on the land; and
the least roughness of the ground hurts their
soles, which are softened by the perpetual
bathing. The water is to them the scene of pleasure and repose; where all their motions are performed with facility, and where their various evolutions are traced with elegance and grace.
View the swans moving sweetly along, or sailing
majestically with expanded plumage upon the
wave! They gaily sport: they dive and again
emerge with gentle undulations, and soft energy;
expressive of those sentiments which are the
foundation of love.

The life of aquatic birds is, therefore, more
peaceful and less laborious than that of most
other tribes. Smaller force is required in swimming than in flying; and the element which they

rather meet with their prey than search for it;

and often a friendly wave conveys it within their reach, and they seize it without trouble or fatigue. Their dispositions are also in general more harmless, and their habits more pacific. Each species congregates, through mutual attachment. They never attack their companions, nor destroy other birds; and in this great and amicable nation, the strong seldom oppress the weak.

'. These birds, in general, have a keen appetite, and are furnished with corresponding weapons. Many species have the inner edges of the bill serrated with sharp indentings, the better to secure their prey: almost all of them are more voracious than the land birds; and there are some, as the ducks and gulls, which devour indiscriminately carrion and entrails.

'Those birds which swim, have palmated and webbed feet; and such as haunt the shores, have divided feet. The latter are differently shaped, their body being slender and tall: and as their feet are not webbed, they cannot dive nor rest on the water; they therefore keep near the brink, and, wading with their tall legs among the shallows, they search, by means of their long neck and bill, for their subsistence among the smaller fish, or in the mud. The amphibious animals occupy the limits between the land and the water, and connect the gradations in the scale of existence.

Mr. Pennant also divides aquatic birds into three orders; those with cloven feet, as the crane kind; those with finned feet, as the snipe kind:

this division, indeed, has some claims to be observed as correct, since those belonging to each part have general and distinct properties; for instance, the waders, or cloven-footed water fowls, are in general tall, light and though with long tails and necks, yet well proportioned; while the web-footed are of a squat make, with a waddling gait; their legs placed far behind, and the length of their necks out of all proportion. Those with finned feet constitute, as it were, a middle race, being calculated both for swimming and wading, and partake of the nature of both. The cloven-footed lay their eggs on the ground, and make no nests. Those with pinnated feet form large nests in the water or near it; and the web-footed fowl deposit their eggs sometimes on lofty cliffs, or inaccessible promontories, or else concealed in the rushes, bushes, &c. near the water. Of the general characteristics of this species a celebrated author has thus ably observed: " The progressions of Nature from one class of beings to another are always by slow and almost

woods and the fields with a variety of the most beautiful birds; and to leave no part of her extensive territories untenanted, she has stocked the waters with its feathered inhabitants also; she has taken the same care in providing for the wants of her animals in this element, as she has

done with respect to those of the other :· she has
used as much precaution to render· water-fowl fit
for swimming, as she did in forming land-fowl
for flight; she has defended their feathers with a
natural oil,· and united their toes by a webbed
membrane; by which contrivances they·have at
once security and ' motion.... But· between, the
classes of land-birds that shun the,water,· and of
water-fowl that ,are made for swimming and
living on it, she bas formed a very numerous
tribe,of birds,' that seem to partake of· a middle
nature ; that, with divided toes, seemingly fitted
to live upon land, are at the same time furnished
with appetites that chiefly attach them to the
waters. These can properly be called ·neither
. land-birds nor water-fowl, as they provide, all
their sustenance from watery ·places, and yet are
unqualified to seek it in those depths where,it is.
often found in greatest plenty.   ·   .(

  " The crane kind are to be distinguished from
others rather by their appetites than their conformation. Yet even in this respect they seem to
be sufficiently discriminated by nature : as· they
·are to live among the.waters, yet are incapable of
swimming in 'them, most of them have long legs,
fitted for wading in shallow waters, or long bills
proper for groping in them.   . , ·ſ)ſſ

  " Every bird of this kind, habituated to marshy
places, may be known, if not by the length of its
legs, at least by the scaly surface of them. Those

who have observed the legs of a snipe or a woodcock, will easily perceive my meaning; and how
different the surface of the skin that covers them
is from that of the pigeon or the partridge.
Most birds of this kind also, are bare of feathers
half way up the thigh; at least, in all of them,
above the knee. Their long habits of wading in
the waters, and having their legs continually in
moisture, prevent the growth of feathers on
those parts; so that there is a surprising difference between the leg of a crane, naked of feathers
almost up to the body, and the falcon, booted almost to the very toes.

"The bill also is very distinguishable in most
of this class. It is, in general, longer than that
of other birds, and in some finely fluted on every
side; while at the point it is possessed of extreme
sensibility, and furnished with nerves, for the
better feeling their food at the bottom of
marshes, where it cannot be seen. Some birds
of this class are thus fitted with every conve-
nience: they have long legs, for wading; long
necks for stooping; long bills, for searching; and
nervous claws, for feeling. Others are not so
amply provided for; as some have long bills, but
legs of no great length; and others have long
necks but very short legs. It is a rule which
universally holds, that where the bird's legs are
long the neck is also long in proportion. It would
indeed be an incurable defect in the bird's

conformation, to be lifted upon stilts' above' its
food, without being furnished with an instrument to reach it.

If we consider the natural power of this class,
in a comparative view, they will seem rather in-
-ferior to those of every other tribe. Their nests
are more simple than those of the sparrow; and
their methods of obtaining food less ingenious
than those of the falcon; the pie exceeds them
in cunning; and though they have the vora-
ciousness of the poultry tribe, they want their
fecundity.—None of this kind, therefore have
been taken into man's society, or under his protection; they are neither caged, like the nightingale; nor kept tame, like the turkey; but lead
a life of precarious liberty, in fens and marshes
at the edges of lakes, and along the sea shore.
They all live upon fish or insects, one or two
only excepted; even those that are called mud-
suckers, such as the snipe and woodcock, it is
more than probable, grope the bottom of marshy
places only for such insects as are deposited
there by their kind, and live in a vermicular state,
in pools and plashes, till they take wing, and become flying insects.

" All this class, therefore, that are fed upon insects, their food being easily digestible, are good
to be eaten; while those who live entirely upon
fish, abounding in oil, acquire in their flesh the
rancidity of their diet, and are, in general, unfit
for our tables. To savages indeed, and sailors

on a long voyage, every thing that has life seems good to be eaten; and we often find them re-commending those animals, as dainties, which they themselves would spurn at, after a course of good living. Nothing is more common in their journals than such accounts as these. ' This day we shot a fox—pretty good eating : this day we shot a heron—pretty good eating : and this day we killed a turtle'—which they rank with the heron and the fox—' as pretty good eating.' Their accounts, therefore, of the flesh of these birds, are not to be depended upon; and when they cry up the heron or the stork of other countries as luxurious food, we must always attend to the state of their appetites, who give the character."

Those who have remarked the feet or toes of a duck, will easily conceive how admirably the web-footed fowl are formed for making way in the water. When men swim they do not open the fingers, so as to let the fluid pass through them; but closing them together present one broad surface to beat back the water, and thus push their bodies along. What man performs by art, nature has supplied to water-fowl; and, by broad skins, has webbed their toes together, so that they expand two broad oars to the water : and thus, moving them alternately with the greatest ease paddle along. We must observe also, that the toes are so contrived, that as they strike backward, their broadest hollow surface beats the water; but as they gather them in again, for a second blow, their front surface contracts, and does not impede the bird's progressive motion.

As their toes are webbed in the most convenient manner, so are their legs also made most fitly for swift progression in the water. The legs of all are short, except three, namely, the flamingo, the avosetta, and the corrira. Except these, all web-footed birds have very short legs; and these strike, while they swim, with great facility. Were the leg long, it would act like a lever whose prop is placed to a disadvantage; its motions would be slow, and the labour of moving considerable. For this reason, the very few birds whose webbed feet are long, never make use of them in swimming; the web at the bottom seems only of service as a broad base, to prevent them from sinking while they walk in the mud; but it otherwise rather retards than advances their motion.

The shortness of the legs in the web-footed kind, renders them as unfit for walking on land, as it qualifies them for swimming in their natural element. Their stay, therefore, upon land, is but short and transitory; and they seldom venture to breed far from the sides of those waters where they usually remain. In their breeding seasons, their young are brought up by the water-side; and they are covered with a warm down, to fit them for the coldness of their situa-

tion. The old ones have a closer, warmer plumage, than birds of any other class. It is of
their feathers that our beds are composed; as
they neither mat nor imbibe humidity, but are
furnished with an animal-oil, that glazes their
surface, and keeps each separate. In some,
however, this animal-oil is in too great abundance; and is as offensive from its smell as it is
serviceable for the purposes of household economy. The feathers, therefore, of all the penguin
kind, are totally useless for domestic purposes;
as neither boiling nor bleaching can divest them
of their oily rancidity. Indeed, the rancidity of
all new feathers, of whatever water fowl they be,
is so disgusting, that our upholsterers give near
double the price for old feathers that they afford
for new: to be free from smell, they must all be
lain upon for some time; and their usual method
is to mix the new and the old together.

The quantity of oil, with which most water
fowl are supplied, contributes also to their warmth
in the moist element where they reside. Their
skin is generally lined with fat; so that, with the

ral lining more internally, they are better defended against the changes or the inclemencies
of the water, than any other class whatever.

Description.

CHAP. V.

" It was a female *Stork* whose mind
Show'd all the mother, bravely kind,
  In trial's fiercest hour:
This bird had bless'd Batavia's lot,
High-nested on a fisher's cot,
  As stedfast as a tow'r."

HAYLEY.

THE STORK.

OF this bird we shall confine ourselves to the most remarkable species which is the white stork, the length of which is about three feet. The bill is nearly eight inches long, and of a fine red colour. The plumage is wholly white; except the orbits of the eyes, which are bare and blackish: some of the feathers on the side of the back and on the wings are black. The skin, the legs, and the bare part of the thighs, are red.

The white stork is semi-domestic: haunting towns and cities; and in many places stalking unconcernedly about the streets, in search of offal and other food. They remove the noxious filth, and clear the fields of serpents and reptiles. On this account they are protected in Holland, and held in high veneration by the Mahomedans; and so greatly respected were they in times of old by the Thessalonians, that to kill one of these birds was a crime expiable only by death.

Bellonius tells us that " the storks visit Egypt in such abundance, that the fields and meadows are white with them. Yet the Egyptians are not displeased with this sight; as frogs are generated in such numbers there, that did not the storks devour them, they would overrun every thing. Besides, they also catch and eat serpents. Between Belba and Gaza, the fields of Palestine are often desert on account of the abundance of mice and rats: and, were they not destroyed, the inhabitants could have no harvest."

The disposition of this bird is mild, neither shy, nor savage: it is an animal easily tamed; and may be trained to reside in gardens, which it will clear of insects and reptiles. It has a grave, air, and a mournful visage: yet, when roused by example, it shews a certain degree of gaiety; for it joins the frolics of children, hopping and playing with them: " I saw in a garden (says Dr. Hermann) where the children were playing at hide-and seek, a tame stork join the party; run in whose turn it was to pursue the rest; so well, as along with the others, to be on its guard." See the annexed engraving.

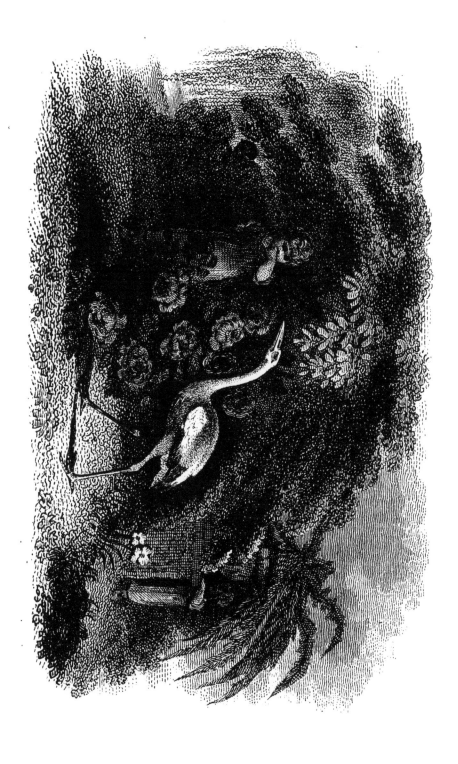

The ancients ascribed many moral virtues to the stork; as temperance, conjugal fidelity, and filial and paternal piety: its manners are such as were likely to attract peculiar attention from them. It bestows much time and care on the education of its young, and does not leave them till they have strength sufficient for defence and support. When they begin to flutter out of the nest, the mother bears them on her wings; she protects them from danger, and will sometimes perish rather than forsake them. A celebrated story is current in Holland: that when the city of Delft was on fire, a female stork in vain attempted several times to carry off her young ones; and finding that she was unable to effect their escape, remained herself in order to share their fate. This extraordinary circumstance furnished Mr. Hayley with the subject of that ballad, from which the motto to the present chapter is extracted.

In "Letters on Italy" is the following anecdote which affords a singular instance of sagacity in this bird. A wild stork was brought by a farmer, in the neighbourhood of Hamburgh, into his poultry-yard, to be the companion of a tame one he had long kept there; but the tame stork disliking a rival, fell upon the poor stranger, and beat him so unmercifully, that he was compelled to take wing, and with some difficulty escaped. About four months afterwards, however, he returned to the poultry-yard, recovered of his

wounds, attended by three other storks who
no sooner alighted than they all together fell
upon the tame stork and killed him.

Storks are birds of passage, and observe great
exactness in the time of their autumnal departure
from Europe to more favourite climates. They
pass a second summer in Egypt and the marshes
of Barbary: in the former country they pair; and
lay again, and educate a second brood. Before
each of their migrations, they rendezvous in
amazing numbers. They are for a while much
in motion among themselves; and after making
several short excursions, as if to try their wings,
all on a sudden take flight with great silence, and
with such speed, as in a moment to be invisible.

During their migrations, they are seen in vast
flocks. Shaw saw three flights of them leaving
Egypt, and passing over Mount Carmel, each
half a mile in breadth; and he says they were
three hours in passing over.

Storks are seldom seen farther north than Sweden: and though they have scarcely ever been
met with in England, they are so common in Holland as to build every where on the tops of the
houses, where the good-natured inhabitants provide boxes for them to make their nests in; and
are careful that the birds suffer no injury, always
resenting this as an offence committed against
themselves. Storks are also common at Aleppo,
and in plenty at Seville, in Spain. At Bagdad,
hundreds of their nests are said to be seen about

5

the houses, walls, and trees ; and at Persepolis,
or Chilmanar, in Persia, the remains of the pil-
lars serve them to build on, " every pillar having
a nest on it," as we read in Fryer's travels.

THE HERON.

THE common heron, which is very frequent in
these kingdoms, is about three feet three inches
in length. The bill is six inches long, and of a
dusky colour. The feathers of the head are long,
and form an elegant crest. The neck is white ;
the fore part marked with a double row of black
spots. The general colour of the plumage is a
blue grey ; with the bastard wing, and greater
quills, black. The middle of the back is almost
bare, and covered by the loose feathers of the
scapulars ; the feathers of the neck also hang
loose over the breast. On each side, under the
wing they are black. The legs are of a dirty
green, and the inner edge of the middle claw is
serrated. The female has no crest, and the feathers on the breast are short.

The different parts in the structure of the heron are admirably adapted to its mode of life. It
has long legs for the purpose of wading : a long
neck, answerable to these, to reach its prey in
the water ; and, a wide throat to swallow it. Its
toes are long, and armed with strong hooked
talons ; one of which is serated on the edge, the

better to retain the fish. The bill is long and sharp; having serratures towards the point, which stand backwards; these, after the prey.is struck,.act like the barbs of a fish-hook, in detaining it till the bird has time to seize it with the claws. Its broad, large, concave, and apparently heavy wings for so small a body, are of great use in enabling it to carry its load to the nest, which is sometimes at a great distance. Dr. Derham tells us, that he has seen lying scattered under the trees of a large heronry, fishes of several inches in length, which must have been conveyed by the birds from the distance of several miles : and D'Acre Barret, Esq. the owner of this heronry, saw a large eel that had been conveyed thither by one of them, notwithstanding the inconvenience that it must have experienced from the fish writhing and twisting about. The body of the heron is very small, and always lean ; and the skin is said to be scarcely thicker than what is called goldbeaters'-skin.

This, of all the birds that are known, is one of the most formidable enemies to the scaly tribe. There is, in fresh waters, scarcely a fish, however large, that he will not strike at and wound, though unable to carry it off: but the smaller fry are his chief subsistence : these, pursued by their larger fellows of the deep, are obliged to take refuge in shallow waters, where they find the heron a still more formidable enemy. His method is to wade as far as he can go into the water, and there patiently wait the approach of his prey; into which, when it comes within his sight, he darts his bill with inevitable aim. Willoughby says he has seen a heron that had no fewer than seventeen carp in his belly at once; these he would digest in six or seven hours, and then go to fishing again. " I have seen a carp (adds that writer) taken out of a heron's belly, nine inches and a half long. Some gentlemen who kept tame herons, to try what quantity one of them would eat in a day, have put several smaller roach and dace in a tub; and they have found him eat fifty in a day, one day with another. In this manner a single heron will destroy fifteen hundred store carp in a single half-year."

This bird, though he usually takes his prey by wading into the water, frequently also catches it while on wing: but this is only in shallow waters, where he is able to dart with more certainty than in the deeps; for in this case, though the fish does, at the first sight of its enemy, descend, yet the heron, with his long bill and legs, instantly pins it to the bottom, and thus seizes it securely. In this manner, after having been seen with its long neck for a minute under water, he will rise upon the wing with a trout or an eel struggling in his bill. The greedy bird, however, flies to the shore, scarcely gives it time to expire, but swallows it whole, and then returns again to his fishing.

Heron-hawking was formerly a favourite diversion in this kingdom; and a penalty of twenty shillings was incurred by any person taking the eggs of this bird. Its flesh was also in former times much esteemed, being valued at an equal rate with that of the peacock.

In seasons of fine weather, the heron can always find a plentiful supply, but in cold and stormy seasons, his prey is no longer within reach; the fish that in the first case come into the shallow water, then keep in the deep, as they find it to be the warmest situation. Frogs and lizards also seldom venture from their lurking places; and the heron is obliged to support himself upon his long habits of patience, and even to take up with the weeds that grow upon the water. At those times he contracts a consumptive disposition, which succeeding plenty is not able to remove; so that the meagre glutton spends his time between want and riot, and feels alternately the extremes of famine and excess. Hence, notwithstanding the care with which he takes his prey, and the amazing quantity he devours, the heron is always lean and emaciated; and though his crop be usually found full, yet his flesh is scarce sufficient to cover the bones.

As this bird does incredible mischief to ponds newly stocked, Willoughby has suggested a method for taking him. " Having found his haunt, get three or four small roach or dace, and having provided a strong hook with a wire to it, this is drawn just within side the shin of the fish, beginning without side the gills, and running it to the tail, by which the fish will not be killed, but continue for five or six days alive. Then having a strong line made of silk and wire, about two yards and a half long, it is tied to a stone at one end, the fish with the hook being suffered to swim about at the other. This being properly disposed in shallow water, the heron will seize upon the fish to its own destruction. From this method we may learn that the fish must be alive, otherwise the heron will not touch them, and that this bird, as well as all those that feed upon fish, must be its own caterer; for they will not prey upon such as die naturally, or are killed by others before them.".

Though this bird live chiefly among pools and marshes, yet its nest is built on the tops of the highest trees, and sometimes on cliffs hanging over the sea. Sometimes as many as eighty have been in one tree. They are never in flocks when they fish, committing their depredations in solitude and silence; but in making their nests they love each other's society; and they are seen, like rooks, building in company with flocks of their kind. Their nests are made of sticks, and lined with wool; and the female lays four or five eggs of a pale green colour. The observable

P 2

indolence of their nature, however, is not less
seen in their nestling than in their habits of depredation. Nothing is more certain than that they
will not be at the trouble of building a nest when
they can get one made by the rook, or deserted
by the owl, already provided for them. This
they usually enlarge and line within, driving off
the original possessors should they happen to renew their claims.

' These birds, if taken young, may be tamed ;
but when the old ones are captured, they soon
pine away, refusing every kind of nourishment. ,

The French avail themselves of the indolence
of this bird, and provide a place with materials
fitted for their nestlings, which they call heron-
ries. The heron, though considered by the
English as unfit for the table, is sought for in
France, where the flesh of the young ones is in
particular estimation. It is therefore for the
purpose of procuring them with more ease, that
they raise up high sheds along some fishy stream;
and furnishing them with materials, the herons
nestle, build, and breed there in great abundance. As soon as the young ones are supposed
to be fit, the owner of the heronry takes, and
carries off such as are proper for eating; and
these are sold for a very good price to the neighbouring gentry. " These are a delicacy which,
(M. Buffon says) the French are very fond of,
but which strangers have not yet been taught to

relish as they ought." Nevertheless it was formerly esteemed as a food in England, and made a favourite dish at great tables. It was then said that the flesh of a heron was a dish-for a king; at present nothing about the house will touch it but a cat.

The herons, therefore, not being considered as worth the trouble of pursuing upon any account whatever, are seldom sought after or disturbed in their retreats, which, excepting when in search after prey, are commonly in almost inaccessible heights. Their nests are often found in great numbers in the middle of large forests, and in some groves nearer home, where the owners have a predilection for the bird, and do not chuse to drive it from their accustomed habitations. . It is certain that by their cries, their expansive wings, their hulk, and wavy motion, they add no small variety to the forest, and solemnity to the scene.

When the young are excluded, as they are numerous, voracious, and importunate, the old ones are for ever upon the wing to provide them with sustenance. The quantity of fish they take upon this occasion is amazing, and their size is not less to be wondered at. Of their assiduity in providing for their young, an instance is given of a heron's nest that was built near a school-house, to which some of the boys climbed up, took down the young ones, sewed up the vent, and laid them in the nest as before. The pain

the poor little animals felt from the operation increased their cries; and this but served to increase the diligence of the old ones in enlarging
their supply. Thus they heaped the nest with
various sorts of fish and the best of their kind;
and as their young screamed they flew off for
more. The boys gathered up the fish, which
the young ones were incapable of eating, till the
old ones at last quitted their nest, and gave up
their brood, whose cravings they found it impossible to satisfy.

The heron is said to be a very long-lived bird;
by Mr. Kepsler's account it may exceed sixty
years; and by a recent instance of one that was
taken in Holland, by an hawk belonging to the
stadtholder, its longevity is again confirmed; the
bird having a silver plate fastened to one leg,
with an inscription, importing, that it had been
struck by the elector of Cologne's hawks thirty-
five years before.

The brown heron has the upper part of the
head, neck, and back, and also the sides of the
wings, of a dark ash-colour; the scapular feathers
have generally white tips with a sort of black
stroke on each side of the wings, the quill-feathers of a more dark colour, very much inclining
to black, except the extreme edges, which are
white; the breast, neck, and upper part of the
belly, are of a pale white, sprinkled with black;
the lower part of the belly darkish ash, and the
thighs of a yellowish cast; the tail is a dark ash,

and the extreme feathers six or seven inches
long.

The blue heron is about the size of the common one, is supposed to weigh upwards of three
pounds, and is about a yard from the tip of the
bill to the end of the toes; the bill is in size and
colour, much the same as the former, only the
upper part is a little hooked at the point. It has
a fine crest of feathers on the top of the head,
which appears of a bluish sky colour; the side
of the head from the bill and under part of the
eyes are white, the covert and scapular feathers
of the wings are of a pale blue, the quill-feathers
black, with their outmost edges blue; the rest of
the body is of a bluish sort of lead colour, it has
yellowish feet, with very long toes, the middle
claw-cerated. This is a curious and very uncommon bird.

John Leo, in his African history, gives an account of a fowl which, by his description, very
much resembles the heron, only its bill, neck,
and legs are somewhat shorter; in flying up, he
says, it mounts out of sight, but descends with a
jirk when it spies a dead carcase; it lives very
long: nay, many of this kind live till age be-
reaves them of all their feathers, upon which they
return to their nest, and are nourished by the
younger birds. They nestle upon high rocks,
and the tops of unfrequented mountains, especially upon mount Atlas, where those who are
acquainted with such places come and take them.

Description—Nests—Food.

The Italians have taken it for a bird of prey, but this author seems of another mind. A Brasilian bird called the soco seems also, in every respect, to resemble the lesser heron.

THE CRANE.

THIS large bird, measures upwards of five feet in length. The bill is above four inches long. The plumage is, in general, ash-coloured; but the forehead is black; and the sides of the head, behind the eyes, and the hind part of the neck, are white; on the upper part of the neck there is a bare ash-coloured space of two inches; and above this the skin is bare and red, with a few scattered hairs. Some parts about the wings

springs an elegant tuft of loose feathers, curled at the ends; which may be erected at will, but which, in a quiescent state, hangs over and covers the tail. The legs are black.

This species is met with in numerous flocks in all the northern parts of Europe. It is said that they make their nests in marshes, and lay two blueish eggs. They feed on reptiles of all kinds, and on some kinds of vegetables; while the corn is green, they are said to make such havock as to ruin the farmers, wherever the flocks alight.

The cranes are migratory; returning northward to breed in the spring, (where they gene-

3

rally make choice of the places which: they occupied the preceding season,) and in ,the winter
inhabiting the warmer regions of Egypt and
India. They fly very high; and arrange themselves in the form of a triangle, the better to
cleave the air. When the wind freshens, and
threatens to break their ranks, they collect their
force into a circle; and they adopt the same disposition when the eagle attacks them. Their
migratory voyages are chiefly performed in the
night; but their loud screams betray their course.
During these nocturnal expeditions the leader
frequently calls to rally his forces, and point out
the track; and the cry is repeated by the flock,
each answering, to give notice that it follows and
keeps its rank.

The flight of the crane is always supported
uniformly, though it is marked by different in-
flections: and these variations have even been
observed to indicate the change of weather; a
sagacity that may well be allowed to a bird,
which, by the vast height to which it soars, is
enabled to perceive the distant alterations and
motions in the atmosphere. Its cry, during the
day, forebodes rain; and its noisy tumultuous
screams announces a storm. If, in a morning
or evening it rise upwards, and fly peacefully in
a body, it is a sign of fine weather; but if it
keep low, or alight on the ground, this menaces
a tempest. When the cranes are assembled on
the ground, they are said to set guards during

the night; and the circumspection of these birds
has been consecrated in the ancient hierogly-,
phies as the symbol of vigilance.

According to Kolben, they are often observed
in large flocks on the marshes about the Cape of
Good Hope. He says, he never saw a flock of
them on the ground that had not some placed,
apparently, as sentinels, to keep a look out, while
the others were feeding; and these, on the approach of danger, immediately gave notice to the
rest. These sentinels stand on one leg; and, at
intervals, stretch out their necks, as if to observe
that all is safe. On notice being given of danger, the whole flock are in an instant on the
wing. Kolben also adds, that in the night-time
each of the watching cranes, which rest on their
left legs, " hold in the right claw a stone of considerable weight; in order that, if overcome by
sleep, the falling of the stone may awake them!"

The crane, like all other large birds (except
the rapacious tribe), has much difficulty in commencing its flight. It runs a few steps; opens
its wings; and then, having a clear space, displays its vigorous and rapid pinions.

Cranes are seen in France in the spring and
autumn; but are, for the most part, merely passengers. It is said that they formerly visited the
marshes of Lincolnshire and Cambridgeshire in
vast flocks, but none have of late been met with.
Their flesh is black, tough, and bad.

THE GIGANTIC CRANE.

THIS is a very large species; measuring, from tip to tip of the wings, nearly fifteen feet. The bill is of a vast size, nearly triangular, and sixteen inches round at the base. The head and neck are naked, except a few straggling curled hairs. The feathers of the back and wings are of a bluish ash-colour, and very stout; those of the breast are long. The craw hangs down the fore part of the neck like a pouch. The belly is covered with a dirty-white down; and the upper part of the back and shoulders are surrounded with the same. The legs and half the thighs are naked; and the naked parts are full three feet in length.

This bird is an inhabitant of Bengal and Calcutta, and is sometimes found on the coast of Guinea. It arrives in the internal parts of Bengal before the period of the rains, and retires as soon as the dry season commences. Its aspect is filthy and disgusting, yet it is one of the most useful birds of these countries, in clearing them of snakes and noxious reptiles and insects. It seems to finish the work begun by the jackal and vulture: they clearing away the flesh of animals, and these birds removing the bones by swallowing them entire. They sometimes feed on fish: and one of them will generally devour as much as would serve four men. On opening the body of a gigantic crane, a land tortoise ten inches long, and a large black male cat, were found entire within it; the former in the craw, and the latter in its stomach. Being altogether undaunted at the sight of mankind, they are soon rendered familiar; and when fish or other food are thrown to them, they catch them very nimbly, and immediately swallow them whole.

The gigantic cranes are believed by the Indians to be animated by the souls of the Brahmins, and consequently to be invulnerable. They are held in the highest veneration .both by the Indians and Africans. Mr. Ives, in attempting to kill some of them with his gun, missed his shot several times; which the by-standers observed with the greatest satisfaction, telling him triumphantly that he might shoot at them as long as he pleased, but he would never be able to kill them.

This very probably is the species mentioned by Mr. Smeathman, as being seen by him in Africa. He describes it as full seven feet high, and appearing at a distance not unlike a " greyheaded man :" on the middle of the neck before was a long conic membrane, like a bladder, covered very sparingly with short down, and rising or falling as the animal moved its beak, but always appearing inflated.

Gigantic cranes are found in companies; and when seen at a distance, near the mouths of rivers, coming towards an observer (which they

Habits—Precautions—Intrepidity.

The bittern is a very retired bird; dwelling chiefly among the reeds and rushes of extensive marshes, where it leads a solitary life, hid equally from the hunter whom it dreads, and the prey that it watches. It frequently continues for whole days about the same spot, and seems to look for safety only in privacy and inaction. In the autumn it changes its abode; always commencing its journey or change of place at sunset. Its precautions for concealment and security seem indeed altogether directed by care and circumspection. It usually sits in the reeds with its head erect; by which means, from the great length of the neck, it sees over their tops, without being itself perceived by the sportsman.

Its principal food during summer consists of fish and frogs; but in the autumn in resorts to to the woods in pursuit of mice, which it seizes with great dexterity, and always swallows whole. About this season it usually becomes very fat.

The bittern, in its general disposition, is not so stupid as the heron, but it is much more ferocious. When caught, it exhibits much rancor, and strikes chiefly at the eyes of its antagonist. Few birds make so cool a defence; it is never itself the aggressor; but, if once attacked, it fights with the greatest intrepidity. If darted on by a bird of prey, it does not attempt to escape; but, with its sharp beak erected, receives the shock on the point, and thus compels its enemy

buzzards never attempt to attack the bittern; and the common falcons always endeavour to rush upon it behind, while it is on the wing.

When wounded by the sportsman, this bird often makes a severe resistance. It does not retire; but waits his onset, and gives such vigorous pushes with his bill, as to wound the leg even through the boot. Sometimes it turns on its back, like the rapacious birds, and fights with both its bill and claws. When surprised by a dog, it is said always to throw itself in this posture. Mr. Markwick once shot a bittern in frosty weather: it fell on the ice, which was just strong enough to support the dogs, and they immediately rushed forward to attack it; but being only wounded, it defended itself so vigorously that the dogs were compelled to leave it, till it was fired at a second time and killed.

During the months of February and March, the males make a kind of deep lowing noise in the mornings and evenings. It is impossible for words to give those who have not heard this evening call an adequate idea of its solemnity. It is like the interrupted bellowing of a bull, but more hollow and louder, and is heard at a mile's distance, as if issued from some formidable being that resided at the bottom of the waters. This extraordinary noise, is produced by a loose membrane, situated at the divarication of the trachæa, capable of great extension, which can be filled

with air and exploded at pleasure. The noise
was formerly believed to be made while the bird
plunged its bill into the mud; hence the poet

> . . . . so that scarce
> The bittern knows his time with bill ingulph'd
> To shake the sounding marsh.

But it has been since discovered, that however,
awful and dismal these bellowing explosions may
seem to us, they are among themselves, the delightful calls to courtship and connubial felicity.
" I remember," says a modern author, " in the
place where I was a boy with what terror this
bird's note affected the whole village; they considered it as the presage of some bad event; and
generally found or made one to succeed it. I do
not speak ludicrously: but if any person in the
neighbourhood died, they supposed it could not
be otherwise, for the night-raven had foretold it;
but if nobody happened to die, the death of a
cow, or a sheep, gave completion to the pro-
phecy."

The nests are formed in April, among rushes;
and almost close to the water, though out of its
reach: they are simple habitations, chiefly composed of the leaves of water-plants and dry
rushes. The female lays four or five greenish
brown eggs, and sits on them for about twenty-
five days. The young, when hatched, are naked
and ugly, appearing almost all legs and neck;

Anecdote—Flavour of its flesh.

they do not venture abroad till about twenty day after extrusion. During this time the parent feed them with snails, small fish or frogs. It i said that the hawks, which plunder the nests o most of the marsh-birds, seldom dare to attac those of the bittern, on account of the old one being always on their guard to defend thei offspring.

We are informed by Latham, that a femal bittern, that was killed during the frost in winte was found to have in her stomach several wart lizards, quite perfect, and the remains of som toads and frogs. These were supposed to hav been taken out of the mud, under shallow wate in the swamp where the bird was shot.

In the reign of Henry the Eighth, the bittei was held in great esteem at the tables of tl great. Its flesh has much the flavour of har and is far from being unpleasant: even now tl poulterers value this bird at about half-a-gu nea. It is consequently much sought for l the fowler, and being a heavy slow-winged bir does not often escape him. Indeed it seldo rises but when almost trod upon, and seems seek protection rather from concealment, the flight. At the latter end of autumn, however, the evening, its wonted indolence appears to fo sake it: It is then seen rising in a spiral asce till it is quite lost from the view, and makes, the same time a singular noise, very differe from its former boomings. The hind-claw, whic is remarkably long, was once supposed a grand preservative for the teeth; and was often set in silver, and used as a tooth-pick.

THE AVOSETTA.

THIS bird is principally found near Milan in Italy; frequently at Rome, and sometimes on the eastern coasts of Suffolk and Norfolk, in the winter. The body is about the size of a pigeon, but very slender made, and tall, being from the tip of the bill to the end of the tail, fourteen inches long, and weighs about nine ounces; its beak is black, flat, and sharp at the end, hooking upwards, which is peculiar to this bird only, about four inches long; the tongue short and not cloven. It has a fine stately pace, or way of walking; its head is not large, but round, and black on the top, and a little way down the back part of the neck; the body entirely white on the under side, the back and covert feathers white, spotted with dusky brown spots; the legs long, of a lovely bright azure colour, bare of feathers above the knees, the claws black, and very small; it has a back toe, which is also small.

From its being bare of feathers above the knee, we may naturally conclude, that it lives by wading in the waters, and that it has also some affinity to the crane kind, by its slender figure; yet, from them it differs in one most essential characteristic, namely, that of being web-footed like the duck. Johnson says, that it has a chirping, pert note; but of its other habits, he gives us not the smallest account, and which, indeed, still remain unknown.

From all the circumstances that have hitherto been collected, the Corrira of Aldrovandus seems to be related to the above; but of this, yet less is known than of the former, and all the information we have is from that author, who says, it has the longest legs of all web-footed fowls, except the flamingo and avosetta; that the bill is straight, yellow, and black at the ends; that the pupils of the eyes are surrounded with two circles, one of which is bay, and the other white: below, near the belly, it is whitish; the tail, with two white feathers, black at the extremities; and that the upper part of the body is of the colour of rusty iron.

THE WATER OUZEL,

CALLED also the water rail, is in size somewhat less than the blackbird. Its bill is black, and almost straight. The eye-lids are white. The upper parts of the head and neck are of a deep brown; and the rest of the upper parts, the

belly, vent, and tail, are black. The chin, the fore-part of the neck, and breast, are white or yellowish. The legs are black.

The water-ouzel frequents the banks of springs and brooks, which it never leaves; preferring the limpid streams whose fall is rapid, and whose bed is broken with stones and fragments of rocks.

The habits of the water-ouzel are very singular. Aquatic birds, with palmated feet, swim or dive; those which inhabit the shores, without wetting their body, wade with their tall legs; but the water-ouzel walks quite into the flood, following the declivity of the ground. It is observed, to enter by degrees, till the water reaches its neck; and it still advances, holding its head not higher than usual, through completely immersed. It continues to walk under the water; and even descends to the bottom, where it saunters as on dry land. M. Herbert communicated to the Comte de Buffon the following account of this extraordinary habit.

" I lay concealed on the verge of the lake Nantua, in a hut formed of pine-branches and snow; where I was waiting till a boat, which was rowing on the lake, should drive some wild ducks to the water's edge. Before me was a small inlet, the bottom of which gently shelved, that might be about two or three feet deep in the middle. A water-ouzel stopped here more than an hour, and I had full leisure to view its manœuvres. It entered into the water, disappeared, and again

emerged on the other side of the inlet, which it
thus repeatedly forded. It traversed the whole
of the bottom, seemed not to have changed its
element, and discovered no hesitation or reluctance in the immersion. However, I perceived several times, that as often as it waded
deeper than the knee, it displayed its wings, and
allowed them to hang to the ground. I remarked
too, that when I could discern it at the bottom of
the water, it appeared enveloped with air, which
gave it a brilliant surface; like some sort of beetles, which in water are always inclosed in a bub-
ble of air. Its view in dropping its wings on entering the water, might be to confine this air; it
was certainly never without some, and it seemed
to quiver. These singular habits were unknown
to all the sportsmen with whom I talked on the
subject; and, perhaps, without the accident of
the snow-hut in which I was concealed, I should
also have for ever remained ignorant of them; but
the above facts I can aver, as the bird came quite
to my feet, and that I might observe it I refrained
from killing it. See the annexed engraving.

Water-ouzels are found in many parts of Europe. The female makes her nest on the ground,
in some mossy bank near the water, of hay and
dried fibres, lining it with dry oak-leaves, and
forming to it a portico or entrance of moss. The
eggs are five in number; white, tinged with a
fine blush of red. A pair of these birds, which
had for many years built under a small wooden

bridge in Caermarthenshire, were found to have a nest early in May : this was taken, but it contained no eggs, although the bird flew out of it at the time. In a fortnight after, they had completed another nest in the same place, inclosing five eggs, which was taken : and in a month after this, a third nest, under the same bridge, was taken, that had in it four eggs ; undoubtedly the work of the same birds, as no others were seen about that part. At the time the last nest was taken, the female was sitting ; and the instant he quitted it she plunged into the water, and disappeared for a considerable time, till at last she emerged at a great distance down the stream. At another time, a nest of the water-ouzel was found in a steep projecting bank (over a rivulet) clothed with moss. The nest was so well adapted to the surrounding materials, that nothing but one of the old birds flying in with a fish in its bill could have led to the discovery. The young were nearly feathered, but incapable of flight ; and the moment the nest was disturbed, they fluttered out and dropped into the water, and, to the astonishment of the persons present, instantly vanished, but in a little time re-appeared at some distance down the stream ; and it was with difficulty that two out of the five were taken.

This bird will sometimes pick up insects at the edge of the water. When disturbed, it usually flirts up its tail, and makes a chirping noise. Its

song in spring is said to be very pretty.   In some
places it is supposed to be migratory.

THE WATER HEN, OR GALLINULE,
THE COOT.

THIS race is considered by naturalists as the
tribe which unites the web-footed kind with those
of the crane species ; for although they have long
legs and necks like the latter, yet by being furnished with a slight membrane between their
toes, they are enabled to swim like the former ; the
principal of them are the water-hen, or gallinule,
and the coots ; these, though placed in different
classes by those who are fond of nice distinctions,
may be said, perfectly to resemble each other
in·figure, feathers, and habits ; they both have
long legs, with thighs almost bare of hair or feathers ; their necks are rather long in proportion ;
their wings short, as are their bills which are very
weak ; their general colour black, and their forehead bald and without feathers.   Such are their
similarities ; and their slight differences are first
in size, the water-hen weighing but fifteen ounces
and the coot twenty-four.   The bald part of the
forehead in the coot is black, in the water-hen it
is of a pink colour.   The toes of the water-hen

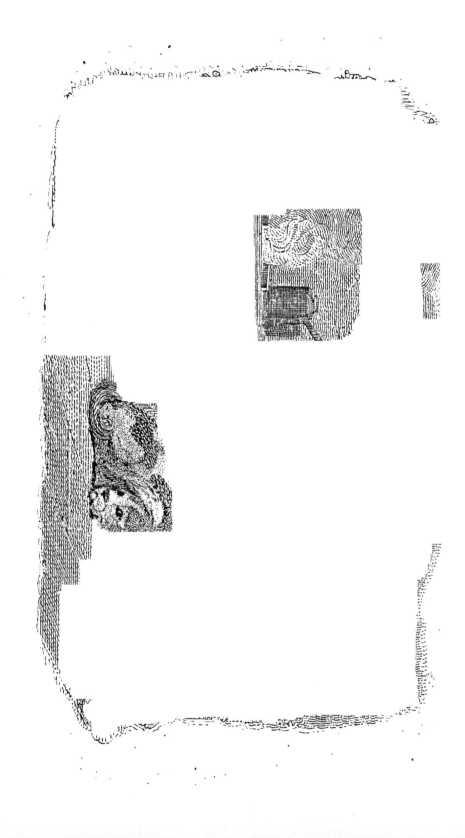

are edged with a straight membrane; those of the coot have it scolloped and broader. In shape and figure their differences are very trifling, and, if possible, in their manner of living, still less; therefore the history of one will serve for, both. As birds of the crane kind are furnished with long wings, and easily change place, the water-hen, whose wings are short, is obliged to reside entirely near those places where her food lies: she cannot take those journies that most of the crane kind are seen to perform; compelled by her natural imperfections, as well perhaps as by inclination, she never leaves the side of the pond or the river in which she seeks for provision. Where the stream is selvaged with edges, or the pond edged with shrubby trees, the water-hen is generally a resident: she seeks her food along the grassy banks; and often along the surface of the water. And it has been judiciously remarked by Goldsmith, that " with Shakespear's Edgar, she drinks the green mantle of the standing pool; or, at least, seems to prefer these places where it is seen. Whether she makes pond-weed her food, or hunts among it for water-insects which are found there in great abundance, is not certain; but I have seen them when pond-weed was taken out of their stomach." She builds her nest upon low trees and shrubs, of sticks and fibres, by the water side. Her eggs are sharp at one end, white, with a tincture of green spotted with red. She lays twice or thrice in a summer; her young ones swim the moment they leave the egg, pursue their parent, and imitate all her manners.

rears, in this manner, two or three broods in a season; and when the young are grown up, she drives them off to shift for themselves.

As the coot is a large bird, it is always seen in larger streams, and more remote from mankind. The water-hen seems to prefer inhabited situations: she keeps near ponds, motes, and pools of water near gentlemen's houses; but the coot keeps in rivers, and among rushy margined lakes. It there makes a nest of such weeds as the stream supplies, and lays them among the reeds, floating on the surface, and rising and sailing with the water. The reeds among which it is built keep it fast, so that it is seldom washed in the middle of the stream. But if this happens, which is sometimes the case, the bird sits in her nest, like a mariner in his boat, and steers with her legs her cargo into the nearest harbour; there, having attained her port, she continues to sit in great tranquillity, regardless of the force of the current; and though the water penetrates her nest, she hatches her eggs in that wet condition.

The water-hen never wanders; but the coot sometimes swims down the current, till it even reaches the sea. In this voyage these birds encounter a thousand dangers; as they cannot fly far, they are hunted by dogs and men; as they never leave the stream, they are attacked and destroyed by otters; they are preyed upon by kites and falcons; and they are taken in still greater numbers, in weirs made for catching fish; for these birds are led into the nets, while pursuing small fish and insects, which are their principal food; and in this instance it has been observed, that " animated nature affords a picture of universal invasion! Man destroys the otter, the otter destroys the coot, the coot feeds upon fish, and fish are universally the tyrants of each other."

THE OYSTER-CATCHER

IS a bird very common on the western shores and sea coasts of South Wales; it generally weighs about half a pound; its bill is of an orange colour, and about two inches and a half long, ending in a sharp point, the upper part being a little longer than the under; the eyes and the edges of their lids are of a fine red; the quill-feathers, head, and all the upper parts of the body down to the middle of the breast are black; except a crescent of white which runs along the throat; the belly, rump, and most of the covert feathers are white; the lower part of the tail is white; but the tips of the tail feathers are all entirely black; the legs and the feet are of a yellowish red, and the middle and outermost toes are united by a slight kind of membrane.

This bird has been called sea-pie, entirely from living on the sea-shores, and its colours being so suddenly contrasted from black to white; and it has been also named oyster-catcher, from the facility with which it takes that fish: for this purpose, whenever it comes near a parcel of them, it patiently watches round until one opens its shells, which is instantly perceived by the bird, who with amazing quickness thrusts in its beak, and almost instantaneously separates the oyster therefrom; besides oysters it feeds upon limpets, and almost all kinds of shell fish; but notwithstanding they make these the principal part of their food, its flesh is rank, and very ill-flavoured.

THE PHALAROPE.

THERE are three varieties. The common one perfectly resembles the sand-piper, except the scolloped membranes on the toes; they are small birds, scarcely ever weighing above an ounce. The grey phalarope has the upper parts of the plumage ash-coloured, varied a little with brown and white, and the breast and belly white. The red phalarope only differs from the former, in having the upper parts of the plumage of a deep lead colour, striped with a dusky yellow, and the under parts of a dusky red. They are not very common birds but are sometimes found in the marshy parts of the country.

THE GREBE.

THIS is larger than either the oyster-catcher or phalarope, and its plumage is white and black; it differs also in the shortness of its legs, which are made for swimming, and not walking; in fact, they are from the knee upward hid in the belly of the bird, and have consequently very little motion. By this mark, and by the scolloped fringe of the toes, may this bird be easily distinguished from all others.

As they are thus, from the shortness of their wings, ill formed for flying, and from the uncommon shortness of their legs, utterly unfitted for walking, they seldom leave the water, but usually frequent those broad shallow pools where the faculty of swimming can be turned to the greatest advantage, in fishing and seeking their prey.

They are chiefly, in this country, seen to frequent the meres of Shropshire and Cheshire; where they breed among reeds and flags, in a floating nest, kept steady by the weeds of the margin. The female is said to be a careful nurse of its young, being observed to feed them most assiduously with small eels; and when the little brood is tired, the mother will carry them, either on her back or under her wings.

This bird preys upon fish, and is almost perpetually diving. It does not shew much more than the head above water; it is very difficult to be shot, as it darts down on the appearance of the

least danger. It is never seen on land; and
though disturbed ever so often, will not leave
that lake where alone by diving and swimming,
it can find food and security. It is chiefly sought
for the skin of the breast, the plumage of which is
of a most beautiful silvery white, and as glossy as
satin. This part is made into tippets; but the
skins are out of season about February, losing their
bright colour; and in breeding time their breasts
are entirely bare.

There are a great number of varieties of the
grebe enumerated, but the most beautiful is the
eared grebe, which is a native of Siberia; it is
about the size of a teal, and is distinguished by a
tuft of orange-coloured feathers, which shoot out

CHAP. VI.

"——————————————The gun
Glanc'd just and sudden, from the fowler's eye,
O'ertakes their sounding pinions; and again
Immediate, brings them from the tow'ring wing,
Dead to the ground; or drives them wide dispers'd,
Wounded, and wheeling various, down the wind."

THOMSON.

THE WOODCOCK.

THIS, commonly termed the snipe genus, is reckoned to include thirty species, of which the woodcock is considered the head. It is not quite so large as the partridge, being from the point of each wing, when extended, about two feet, and weighs about eleven, but sometimes twelve ounces; the bill is straight, and about three inches long, the upper part falling a little over the under at the tip end; the back, and all the under parts of the body partake of a great variety of colours; the back part of the head inclining to black, with little cross bars that appear like a sort of shell work; and between the eye and the bill, a black line on each side ; nearer to the bill, it is more reddish, the whole beautifully variegated with red, black, grey, and ash-colour, which viewed together makes a very delightful appearance; the breast and belly are more grey, with a variety of transverse pale and brown lines. The sides of the wings are crossed with various red bars, like those on the head, with a few pale or whiter feathers, interspersed upon each; the under parts of the wings are a mixture of grey and brown, with a variety of crossed or waved lines. The tail is about three inches long, the upper part of the tips cinereous or brown, the under white, which when it raises its tail, as it frequently does while feeding, is often discovered by those feathers. The legs and feet are of a dusky pale colour; and the claws, which are very small, black.

Woodcocks, according to their name, frequent woods, and woody places, where there are rivulets; they are very often found also on the sides of banks, near watry ditches, and in small brambles and coverts, where they feed amongst slime and earth ; from whence Mr. Willoughby says they draw small shell fish, worms, and other insects ; but Mr. Durham is of opinion that they feed chiefly on the fatty unctuous humour they suck out of the earth, for which purpose he says they have remarkable nerves reaching to the end of their bills, peculiarly appropriated to the purposes of insertion and suction. They go out in the evening, feeding, and flying, principally in the night, and generally return in the same direction, or through the same glades, to their day retreat.

These birds of passage during summer, inhabit Norway, Sweden, Lapland and other northern countries, where they breed. As soon, however, as the frosts commence they retire southward to milder climates. They arrive in Great Britain in flocks, some of them in October, but not in great numbers till November and December. They generally take advantage of the night, being seldom seen to come before sun-set. The time of their arrival depends considerably on the prevailing winds; for adverse gales always detain them; not being able to struggle with the boisterous squalls of the northern ocean. After their arrival in bad weather, they have often been seen so much exhausted as to allow themselves to be seized by the hand when they alighted near the coast.

The greater part of the woodcocks leave this country about the latter end of February or the beginning of March, always pairing before they set out. They retire to the coast, and if the wind be fair, set out immediately; but if contrary, they are often detained in the neighbouring woods and thickets for some time. In this crisis the sportsmen are alert, and the whole surrounding country echoes the discharge of guns : seventeen brace have been killed by one person in a day. But if they are detained long on the dry heaths, they become so lean as to be scarcely eatable. The instant a fair wind springs up, they seize the opportunity; and where the sportsman has seen hundreds in one day, he will not find even a single bird the next.

This bird, being a very clumsy waddling walker, (as is the case with every kind of fowl having short legs and long wings) when flushed, rises heavily from the ground, and makes a considerable noise before he can gather wind sufficient for flight. If found in a rushy spot, a ditch, or a hedge-row, from whence he is obliged to present an *open* mark, he frequently slowly skims over the ground, and is very easily shot, as, indeed, is the case elsewhere, provided only obstruction do not arise from intervening branches of trees and boughs of underwood, which in cock and covert shooting, must always be expected. Woodcocks may be found as well with pointers as with spaniels, (the pointers being hunted in the covert with bells) but cock-shooting with spaniels is almost universally preferred as it is more enlivening to hear the spaniels occasionally in *quest,* than to pursue so pleasing a scene with the solemnity of silence.

Very few woodcocks breed in England; and perhaps in those that do, it may be owing to their having been so wounded by the sportsmen in the winter, as to be disabled from taking their long journey in spring. They build their nests on the ground, generally at the root of some tree; and lay, four or five eggs, about the size of those of a pigeon, of a rusty colour, and marked with brown spots. They are remarkably tame during incubation: a person who discovered a woodcock on its nest, often stood over, and even stroked it; notwithstanding which, it hatched the young, and in due time disappeared with them.

A single bird was observed to remain in a coppice belonging to a gentleman in Dorsetshire through the summer. The place, from its shady and moist situation, was well calculated to maintain it; yet by degrees it lost almost all its feathers, so that for some time it was not able to fly, and was often caught; but in the autumn it recovered its feathers and strength, and flew away.

It has been remarked in England, that for several years past, woodcocks have become very scarce. This seems to be easily accounted for. Sweden, like other countries, is making a gradual progress in the arts of luxury; among which the indulgence of the palate fills no undistinguished place. The eggs of wild-fowl have of late become a great delicacy among the inhabitants of that country, who encourage the boors to find out their nests. The eggs of the woodcock they are particularly fond of; and the boors offer them in large quantities for sale, in the market of Stockholm. From this practice it is not improbable that the breed, not only of this bird, but of several of the species of grous, will be greatly diminished, if not at last totally extirpated.

T 2

The inhabitants of the North of Europe, to whose forests the woodcocks retire in the summer, never eat them; esteeming their flesh unwholesome, from the circumstance of their having no crops.

In Lancashire, great numbers of woodcocks are taken in traps in moonlight nights. Long parallel rows of stones and sticks, about four or five inches, are made on the commons which they frequent. In these rows several intervals or gateways are left, in which the traps are placed. When the bird, running about in search of food, comes to one of these rows, he will not cross it, but runs along the side till he comes to a gateway; which he enters, and is then taken.

THE GODWIT

IS about sixteen inches in length, and weighs from ten to twelve ounces; its bill is near as long as that of the woodcock, of a palish red towards the base, and black at the point, the upper mandible something longer than the lower, the tongue is sharp, the ears open, and large.

The feathers upon the head are of a light brown or reddish colour, with their middle parts black, but about the eyes of a more pale or yellowish tincture; the neck and breast are pretty much of the same colour with the head, only interspersed with transversed black lines, edged with a pale yellow.

Description.

The large wing feathers are black, the shafts white, with a broad bar of white running along the middle of the three first feathers; the rest of the row, and those also of the next have reddish ash-coloured edges and tipt: the lesser covert feathers are of the same colour as the body. The tail feathers are alternately crossed with black and white lines. The legs of a dusky greenish colour, and the claws black.

They feed by the sea-side upon sandy shores; down like the gull. The throat and neck of the hen are grey, and the rump white, speckled or powdered with blackish spots. They are in some places called the stone plover.

THE GREEN-SHANK.

THIS is not so common as the godwit; it is about fourteen inches in length; the bill two inches and a half long. The plumage on the upper parts is a brown ash colour; on the lower parts white; and it has a broad white stroke extending from the bill to the eye: the legs are green, whence it takes its name: It has the same manners and character as the godwit, and has also a white line over the eye: but does not weigh more than half as much.

THE RED-SHANK

WEIGHS about five ounces and an half, and is twelve inches long. The bill is two inches, red at the base, and black towards the point. The head, neck, and scapulars are dusky ash-colour, obscurely spotted with black : the back is white, spotted with black : the breast is white, streaked with dusky lines. When its nest is in danger, it makes a noise somewhat similar to that of the lapwing.

THE SNIPE,

FROM the point of the bill to the end of the tail, is about twelve inches, and from the point of each wing when extended about fifteen or sixteen, the head is divided by a pale and red line, which runs longways, parallel to which on each side is a black line, and over the eyes there runs another line pretty much of the same colour of

place under the bill. The feathers that spring from the shoulders are so long that they reach almost as far as the end of the tail, the outward half from the shaft being of a pale red. The colours thus succeeding each other, make two

lines down the back, the covert feathers of which
are dusky with white transverse lines, and white
tips on some of the large wing feathers, the lesser
feathers being of a mixed colour of red, black,
and grey, beautifully variegated with white and
brown lines: the tail feathers are more red, with
black lines running across them. The bill is
black at the tip, and near three inches long, the
tongue is sharp, the eyes of a hazel colour. The
legs are of a pale greenish colour, the toes pretty
long and the talons black.

There are two sorts, but they frequent the
same places, subsist on the same food and are
frequently found near to each other. The larger
is called the whole snipe, and the smaller the
jack.

The flesh is exceedingly good, sweet, and tender; it feeds in drains of water springs, and other
fenny places, on worms and other insects, and
upon the fat unctious humour that it sucks out
of the earth.

Snipes are well known to the sporting world
in winter shooting; a jack snipe is not very
easily killed, at least by an indifferent shot, of
which some proof was recently given by a gentleman of Easthampstead, in Windsor Forest,
who very warmly entertained his friend with a
description of a jack snipe he had found upon
the heath, which had afforded him sport for *six
weeks;* and he did not at all doubt but he would

serve him for sport during the season, if he were. not *taken off* by frost; and what was still more *convenient,* he always knew where to find him within a hundred yards of the place,

Snipes are birds of passage, supposed to breed principally in the lower lands of Switzerland and Germany; though some (particularly the jacks) remain and breed in the fens, and marshy swamps of this country, where their nests and eggs are frequently found. They lay four or five eggs. They arrive here sooner or later in the autumn, regulated in respect to time, by the wind and weather; but never appear till after the first rains, and leave this country in the spring, as soon as the warmer sun begins to absorb or exhale the moisture from the earth, and denote the approach of summer.

THE SANDPIPER.

THERE are at least forty varieties of this genus; among which, besides the two following articles, are the knot, the puno, the turnstone, and the dunlin.

The sandpiper is a small bird, seldom exceeding the size of a thrush, at least in England, and some of them are not bigger than a sparrow. In the milder climates there are larger species, such as the green, the spotted, the red, and the gam-

Description.

bol sandpipers, many of which have been seen as large as pigeons.

The sandpiper of England weighs about two ounces; it has a brown head, streaked with black, the back and coverts, brown, mixed with a glossy green, and the breast and belly quite white. The bill is straight and slender, about an inch and a half long; the nostrils small, and the tongue slender. The toes are divided or slightly connected at the base by a membrane; the hinder toe is short and weak.

The whole of this tribe have a shrill pipe, or whistle, from which they derive their name, and which they constantly make use of.

THE RUFF AND REEVE.

THE ruff, which is of the sandpiper tribe, is about a foot in length, with a bill of about an inch. The face is covered with yellow pimples; and the back part of the head and neck are furnished with long feathers, standing out somewhat like the ruff worn by our ancestors; a few of these feathers stand up over each eye, and appear not unlike ears. The colours of the ruffs are in no two birds alike: in general they are brownish, and barred with black; though some have been seen that were altogether white. The lower parts of the belly and the tail coverts are white. The tail is tolerably long, having the four middle feathers barred with black; the others are pale brown. The legs are of a dull yellow, and the claws black. The female, which is called the reeve, is smaller than the male, of a brown colour, and destitute of the ruff on the neck.

The male bird does not acquire his ruff till the second season, being till that time in this respect like the female: as he is also from the end of June till the pairing season, when nature clothes him with the ruff, and the red pimples break out on his face; but after the time of incubation the long feathers fall off, and the caruncles shrink in under the skin so as not to be discerned.

These are birds of passage; and arrive in the fens of Lincolnshire, the Isle of Ely, and the East Riding of Yorkshire, in the spring, in great numbers. It is not known with certainty in what countries they pass their winter. Mr. Pennant tells us, that in the course of a single morning there have been above six dozen caught in one net: and that a fowler has been known to catch between forty and fifty dozen in a season.

The ruffs are much more numerous than the reeves, and they have many severe contentions for their mates. The male chuses a stand on some dry bank, near a splash of water, round which he runs so often as to make a bare circular path: the moment a female comes in sight, all the

3

Contentious spirit.

males within a certain distance commence a general battle; placing their bills to the ground, spreading their ruff, and using the same action as a cock : and this opportunity is seized by the fowlers, who, in the confusion, catch them, by means of nets, in great numbers; yet even in captivity, their animosity still continues.

This bird is so noted for its contentious spirit, that it has obtained the epithet of the fighter. In the beginning of spring, when these birds arrive among our marshes, they are also observed to engage with desperate fury against each other.

An erroneous opinion prevails very generally, that ruffs, when in confinement, must be fed in the dark, lest the admission of light should set them to fighting. The fact is, that every bird, even when kept in a room, takes its stand, as it would in the open air; and if another invade its circle, a battle ensues. A whole roomful of them may be set into fierce contest by compelling them to shift their stations; but after the disturber has quitted the place, they have been observed to resume their circles, and become again pacific. In confinement their quarrels originate in the circumstance of the pan containing their food not being sufficiently large to admit the whole party to feed without touching each other. When the food has been divided into several pans, the birds have continued perfectly quiet.

The reeves lay four eggs, in a tuft of grass, about the beginning of May; and the young are hatched in about a month.

THE LAPWING, OR PEE-WIT.

THIS bird is about the size of a common pigeon, and is covered very thick with plumes, which are black at the roots, but of a different colour on the outward part. The feathers on the belly, thighs, and under the wings, are most of them white as snow; and the under part on the outside of the wings white, but black lower. It has a great liver, divided into two parts, and, as some authors affirm, no gall.

Lapwings are found in most parts of Europe, as far northward as Iceland. In the winter they are met with in Persia and Egypt. Their chief food is worms; and sometimes they may be seen in flocks nearly covering the low marshy grounds in search of these, which they draw with great dexterity from their holes. When the bird meets with one of these little clusters of pellets, or rolls of earth, that are thrown out by the worm's perforations; it first gently removes the mould from the mouth of the hole, then strikes the ground at the side with its foot, and steadily and attentively waits the issue: the reptile, alarmed by the shock, emerges from its retreat,

and is instantly seized. " To ascertain this circumstance," says M. Baillon, " I employed the
same stratagem: in a field of green corn, and in
the garden, I beat the earth for a short time, and
I saw the worms coming out. I pressed down a
stake, which I then turned in all directions to
shake the soil; this method succeeded still
quicker; the worms crawled out in crowds, even
at the distance of a fathom from the stake." In
the evening the lapwings pursue a different plan:
they run along the grass, and feel under their
feet the worms, which now come forth invited
by the coolness of the air. Thus they obtain a
plentiful meal; and afterwards wash their bill
and feet in the small pools or rivulets.

" I have seen this bird," says Dr. Latham,
" approach a worm-cast, turn it aside, and, after
making two or three turns about, by way of giving motion to the ground, the worm came out,
and the watchful bird, seizing hold of it, drew it
forth."

These birds make a great noise with their
wings in flying, and are called pee-wits in the
North of England, from their particular cry.
They remain here the whole year. The female
lays two eggs on the dry ground, near some
marsh; upon a little bed which she prepares of
dry grass. These are olive-coloured, and spotted
with black. She sits about three weeks; and
the young, who are covered with a thick down,

are able to run within two or three days after they are hatched.

The parent exhibits the greatest attachment to them; and the arts used by this bird to allure boys and dogs from the place where. they are running, are extremely amusing. She does not wait the arrival of her enemies at the nest, but boldly pushes out to meet them. When as near as she dare venture, she rises from the ground with a loud screaming voice, as if just flushed from hatching, though probably at the same time not within a hundred yards of her nest. She now flies with great clamour and apparent anxiety; whining and screaming round the invaders, striking at them with her wings, and sometimes fluttering as if she was wounded. To complete the deception, she becomes still more clamorous as she retires from the nest. If very near, she appears altogether unconcerned; and her cries cease in proportion as her fears are augmented. When approached by dogs, she flies heavily, at a little distance before them, as if maimed; still vociferous, and still bold, but never offering to move towards the quarter where her young are stationed. The dogs pursue in expectation every moment of seizing the parent, and by this means actually lose the young; for the cunning bird, having thus drawn them off to a proper distance, exerts her powers, and leaves her astonished pursuers to gaze at the rapidity of her flight.

The following anecdote, which was communicated to Mr. Bewick by the Rev. J. Carlyle, and is here illustrated by an engraving, exhibits the domestic nature of the lapwing; as well as the art with which it conciliates the regard of animals materially differing from itself, and generally considered as hostile to every species of the feathered tribe. Two lapwings were given to a clergyman, who put them into his garden; one soon died, but the other continued to pick up such food as the place afforded, till winter deprived it of its usual supply. Necessity soon compelled it to draw nearer the house; by which it gradually became familiarized to occasional interruptions from the family. At length one of the servants, when she had occasion to go into the back-kitchen with a light, observed that the lapwing always uttered his cry of *"pee-wit"* to obtain admittance. He soon grew more familiar: as the winter advanced, he approached as far as the kitchen; but with much caution, as that part of the house was generally occupied by a dog and a cat, whose friendship, however, the lapwing at length conciliated so entirely, that it was his regular custom to resort to the fireside as soon as it grew dark, and spend the evening and night with his two associates, sitting close by them, and partaking of the comforts of the warmth. As soon as spring appeared, he discontinued his visits to the house, and betook himself to the garden; but on the approach of winter he

received him very cordially. Security was pro-

with caution, was afterwards taken without re-
serve: he frequently amused himself with washing
in the bowl which was set for the dog to drink
out of; and while he was thus employed, he
shewed marks of the greatest indignation if either
of his companions presumed to interrupt him.
He died in the asylum he had thus chosen, being
choaked with something that he picked up from
the floor.

These birds are accounted very delicate eating,
the flesh being tender and well tasted.

THE DOTTEREL.

The dotterel is about ten inches in length, and
weighs about four ounces. The bill is not quite
an inch long, and is black. The forehead is
mottled with brown and grey: the top of the
head is black; and over each eye there is an
arched line of white, which passes to the hind

white; the back and wings are of a light brown
inclining to olive, each feather margined with
pale rust colour. The fore part of the neck is
surrounded by a broad band of a light olive colour, bordered below with white. The breast is
of a pale dull orange; the middle of the belly

black; and the rest of the belly and the thighs are of a reddish white. The tail is olive brown, black near the end, and tipped with white; and the outer feathers are margined with white. The legs are of a dark olive. The colours of the female are less vivid.

These are migratory birds: appearing in flocks of eight or ten, about the end of April; and staying all May and June, when they become very fat, and are much esteemed for the table. They are found in tolerable plenty in Cambridgeshire, Lincolnshire, and Derbyshire; but in other parts of the kingdom they are scarcely known. They are supposed to breed among the mountains of Westmoreland and Cumberland.

The dotterel is not only a very singular bird in its manner, but also very foolish, as it may be taken by the most simple artifice. The country people are said sometimes to go in quest of it, in the night, with a lighted torch or candle: and the bird, on these occasions, will mimic the actions of the fowler with great archness. When he stretches out an arm, it stretches out its wing; if he move a foot, it moves one also; and every other motion it endeavours to imitate. This is the opportunity that the fowler takes of entangling it in his net. Willoughby, however, cites the following case: " Six or seven persons usually went in company to catch dotterels. When they found the bird, they set their nets in an advantageous place: and each of them holding a stone in either hand, they got behind it, and striking the stones often one against the other, roused it from its natural sluggishness, and by degrees drove it into the net." The more certain method of the gun has of late nearly superseded both these artifices.

THE LONG-LEGGED PLOVER.

THE following very pleasing description of this bird is given by Mr. White.

" In the last week of April 1779, five of these most rare birds (which are too uncommon to have an English name, but are known to naturalists by the terms himantopus, or loripes, or charadrius himantopus) were shot upon the verge of Frensham-pond; a large lake belonging to the bishop of Winchester, and lying between Woolmer-forest and the town of Farnham, in the county of Surrey. The pond-keeper says there were three brace in the flock; but that after he had satisfied his curiosity, he suffered the sixth bird to remain unmolested.

" One of these specimens I procured; and found the length of the legs to be so extraordinary, that at first sight one might have supposed the shanks had been fastened on, to impose on the beholder: they were legs *in caracatura;* and

5

Uncommon length of its legs.

had we seen such proportions on a Chinese or
Japan screen, we should have made large allowance for the *fancy* of the draughtsman.

"These birds are of the plover family, and
might with propriety be called the stilt-plovers.
My specimen, when drawn and stuffed with pep-
per, weighed only four ounces and a quarter,
though the naked part of the thigh measured
three inches and a half. Hence we may safely
assert, that these birds exhibit weight for inches,
and have incomparably the greatest length of legs
of any known bird. The Flamingo, for instance,
is one of the most long-legged birds, and yet it
bears no manner of proportion to the himantopus: for a cock flamingo weighs, at an average,
about four hundred pounds avoirdupois; and
his legs and thighs measure usually about twenty
inches. But four pounds are fifteen times and a
fraction more than four ounces and a quarter; and
if four ounces and a quarter have eight inches of
legs, four pounds must have one hundred and
twenty inches and a fraction of legs, or somewhat more than ten-feet; such a monstrous proportion as the world never saw!" Here the Rev.
Mr. Bingley remarks, that Mr. White appears
to have calculated the weights of these birds unfairly; the plover after it was stuffed, and the
flamingo from a perfect bird; which, in the comparison of weights, will make a difference extremely material.

"If," continues Mr. White, "we try the ex-

x 2

periment in still larger birds, the disparity would increase. It must be matter of great curiosity to see the stilt plover move; to observe how it can wield such a length of lever with such feeble muscles as the thighs seem to be furnished with. At best, one should expect it to be but a bad walker : but what adds to the wonder is, that it has no back toe. Now, without that steady prop to support its steps, it must, theoretically, be liable to perpetual vacillations, and seldom able to preserve the true centre of gravity.

" These long-legged plovers are birds of South Europe, and rarely visit our island; and when they do, are wanderers and stragglers, and impelled to make so distant and northern an excursion from motives or accidents for which we are not able to account."

This bird is common in Egypt and the warmer parts of America, where it feeds on flies and other insects; but it is very rare in England.

THE GREEN PLOVER

IS much about the same size as the lapwing, and has a short round black bill, sharp at the end, and a little hooked. The tongue, which fills all the lower chap of the bill, is triangular at the tip, horny underneath, and turns a little up. The feathers of the back and wings are black, thick set with transverse spots of a yellowish

Where found—Food—Description.

green colour;, the breast is brown,, spotted with yellowish green ; and the belly white. It has no hind claw or spur.

These birds are found in France; Switzerland, Italy, and in most countries of England in all which places they are esteemed a choice dish, their flesh being very tender, and of an exceeding agreeable flavour. They feed chiefly upon worms ; though some authors have affirmed they live, like the grasshopper, upon nothing but dew.

This bird was called paradalis by the ancients, from its beautiful spots, which somewhat resemble those of the leopard.

THE GREY PLOVER

IS about the bigness of the former ; but the bill is somewhat longer and thicker, and it has a very small hind claw or spur. The head, back, and lesser feathers of the wings, are black, with tips of a greenish grey. The breast, belly, and thighs are white, as are also the feathers under the bill ; and the throat is spotted with brown or dusky spots. The tail is very short, insomuch that the wings exceed it in length.

The flesh of this bird is very tender and delicate, and no less esteemed than that of the former.

weighs only two ounces; the bill is half an inch long, and from it to the eyes runs a black line. The upper part of the neck is encircled with a white collar, the lower part with a black one. The back and wings are light brown, the breast and belly are white, the legs yellow. They frequent our shores in summer.

THE KNOT.

THIS bird measures not more than nine inches and weighs only four ounces and an half. The head and neck are ash-colour, the back and scapulars brown, with a white bar on the wings. They frequent the coasts of Lincolnshire from August to November, and when fattened, are preferred by some to the ruffs themselves.

THE PURRO

WEIGHS only an ounce an half, and is in length seven inches. A white stroke divides the bill and eyes. The upper parts of the plumage are brownish ash-colour, and the breast and belly white, as are the lower parts of the quill-feathers.

These birds come in vast flocks on our sea-coasts in winter, and in their flight observe uncommon regularity, appearing like a white or dusky cloud. They were formerly a frequent dish at our tables and known by the name of stints.

THE TURNSTONE

IS about the size of a thrush. The bill is nearly an inch long, and turns a little upwards. The head, throat and belly are white, the breast black; and the neck encircled with a black colour. The upper parts of the plumage are of a pale reddish brown. These birds take their name from their method of finding their food, which is by turning up small stones with their bills, to get at the insects which lurk under them.

THE DUNLIN.

THIS is the size of a jack snipe. The upper parts of the plumage are ferruginous, marked with large spots of black, and a little white; the lower parts are white, with dusky streaks. It is found in all the northern parts of Europe.

THE BALEARIC CRANE.

THIS bird presents to the eye a very whimsical figure. It is pretty nearly the shape and size of the ordinary crane, with long legs and a long neck, like others of the kind; but the bill is shorter, and the colour of the feathers of a dark

greenish grey. The head and throat form the
most striking part of this bird's figure. On the
head is seen standing up, a thick round crest,
made of bristles, spreading every way, and resembling rays standing out in different directions.
The longest of these rays are about three inches
and an half; and they are all topped with a kind
of black tassels, which give them a beautiful appearance. The sides of the head and cheeks are
bare, whitish, and edged with red, while underneath the throat hangs a kind of bag or wattle,
like that of a cock, but not divided into two. To
give this odd composition a higher finishing, the
eye is large and staring; the pupil black and big,
surrounded with a gold-coloured iris, that com-
pletes the bird's very singular appearance.

These birds come from the coast of Africa, and
the Cape de Verd Islands. As they run, they stretch
out their wings and go very swiftly; otherwise their
usual motion is very slow. In their domestic
state, they walk very deliberately among other
poultry, and suffer themselves to be approached
by every spectator. They never roost in houses;
but about night, when they are disposed to go to
rest, they search out some high wall, on which
they perch in the manner of a peacock. Indeed
they resemble that bird very much in manners
and disposition. But, though their voice and
roosting be similar, their food, which is entirely
upon greens, vegetables, and barley, seems to
make some difference.

THE JABIRU AND JABIRU GUACU.

THESE are both birds of the crane kind, and natives of Brasil.; The bill of the latter is red, and thirteen inches long; the bill of the former is black, and is found to be eleven. Neither of them, however, are of a size proportioned to their immoderate length of bill. The jabiru guacu is not above the size of a common stork, while the jabiru with the smallest bill exceeds the size of a swan. They are both covered with white feathers, except the head and neck, which are naked: and their principal difference is in the size of the body and the make of the bill; the lower chap of the jabiru guacu being broad, and bending upwards.

THE ANHIMA.

THIS is a water-fowl of the rapacious kind, and bigger than a swan. The head, which is small for the size of the body, bears a black bill, which is not above two inches long; but what distinguishes it in particular is a horn growing from the forehead as long as the bill; and bending forward like that of the fabulous unicorn of the ancients. This horn is not much thicker than a crow-quill, as round as if it were turned in a lathe, and of an ivory colour. But this is

nqt the only instrument of battle this formidable
bird carries; it seems to be armed at all points;
for at the forepart of each wing, and the second
joint, spring two straight triangular spurs, about
as thick as one's little finger: the foremost of
these goads or spurs is above an inch long; the
hinder is shorter, and both of a dusky colour.
The claws are also long and sharp; the colour is
black and white; and they cry terribly loud,
sounding something like "vyhoo, vyhoo." They
are never found alone, but always in pairs; the
cock and hen prowl together; and their fidelity
is said to be such, that when one dies, the other
never departs from the carcase, but dies with its
companion. It makes its nest of clay, near the
bodies of trees, upon the ground, of the shape of
an oven. This bird is also a native of Brasil.

THE NUMIDIAN CRANE,

WHICH from the peculiarity of its manners,
is vulgarly called by our sailors the buffoon bird,
and by the French, the mademoiselle, or lady.
The same qualities have procured it these different appellations from two nations, who, on more
occasions than this, look upon the same objects
in very different lights. The peculiar gestures
and contortions of this bird, are extremely singular; and the French, who are skilled in the arts
of elegant gesticulation, consider all its motions

as ladylike and graceful. Our English sailors, however, who have not entered so deeply into the dancing art, think, that while thus in motion, the bird cuts but a very ridiculous figure. It stoops, rises, lifts one wing, then another, turns round, sails forward, then back again; all which highly diverts our seamen; not imagining, perhaps, that all these contortions are but the aukward expressions, not of the poor animal's pleasures, but its fears.

It is a very scarce bird; the plumage is of a leaden grey; but it is distinguished by fine white feathers, consisting of long fibres, which fall from the back of the head, about four inches long; while the fore part of the neck is adorned with black feathers, composed of very fine, soft, and long fibres, that hang down upon the stomach, and give the bird a very graceful appearance. It comes from that country from whence it has taken its name. The ancients have described a buffoon bird, but there are many reasons to believe that theirs is not the Numidian crane.

THE SPOONBILL.

THE European spoonbill is about the bulk of a crane; but the latter is above four feet high, while the former is seldom more than three. The common colour of those of Europe is a dirty white; but those of America are of a beautiful

rose-colour, or a delightful crimson. Beauty of
plumage seems to be the prerogative of all the
birds of that continent; and we here see the
most splendid tints bestowed on a bird, whose
figure is sufficient to destroy the effects of its colouring; for its bill is so oddly fashioned, and its
eyes so stupidly staring, that its fine feathers only
tend to add splendor to deformity. The bill,
which in this bird is so very particular, is about
seven inebes long, running out broad at the
end, as its name justly serves to denote; it is of
a red colour, like the rest of the body. All round
the upper chap there runs a kind of rim, with
which it covers that beneath; and as for the rest,
its cheeks and its throat, are without feathers,
and covered with a black skin.

This bird seems to lead a life entirely like that
of the crane kind; its toes are divided, and it
feeds among waters upon frogs, toads, and serpents; of which, particularly at the Cape of
Good Hope, they destroy great numbers. The
inhabitants of that country hold them in as much
esteem as the ancient Egyptians did their bird
ibis: it runs tamely about their houses; and they
are content with its society, as an useful though
an homely companion. They are never killed;
and indeed they are good for nothing when they
are dead, for the flesh is unfit to be eaten.

These birds breed in Europe, in company with
the heron, in high trees; and in a nest formed
of the same materials. They lay from three to

five eggs, white, and sprinkled with a few san-
guine or pale spots. Willoughby tells us, that
in a certain grove, at a village called Seven
Huys, near Leyden, they build and breed yearly
in great numbers. In this grove also, the heron,
the bittern, the cormorant, and the shag, have
taken up their residence, and annually bring forth
their young together. Here the crane kind
seems to have formed their general rendezvous;
and, as the inhabitants say, every sort of bird
has its several quarter, where none but their own
tribe are permitted to reside. Of this grove the
peasants of the country make good profit. When
the young ones are ripe, those that farm the
grove, with a hook at the end of a long pole,
catch hold of the bough on which the nest is
built, and shake out the young ones; but sometimes the nest and all tumble down together.

THE FLAMINGO.

NOTWITHSTANDING the flamingo is
web-footed like birds of the goose kind, its
height, figure, and appetites render it more of
the crane species. With a longer neck and legs
than any other of the crane kind, it seeks its
food by wading among waters, and only differs
from all of this tribe in the manner of seizing its
prey; the heron makes use of its claws, but the
flamingo uses only its bill, which is strong and

2

thick for the purpose, the claws being useless, as they are feeble, and webbed.

This bird is the most remarkable of all the crane kind, being not only the tallest, but the most bulky and handsome. The body, which is of a beautiful scarlet, is not bigger than that of a swan; but its legs and neck are of such an extraordinary length, that when it stands erect, it is upwards of six feet high. Its wings, extended, are five feet six inches from tip to tip; and it is four feet eight inches from beak to tail. The head is round and small, with a large bill, seven inches long; partly red, partly black, and crooked like a bow. The legs and thighs, which are not much thicker than a man's finger, are about two feet eight inches high; and its neck near three feet long. The feet (as before observed) are not furnished with sharp claws, as in others of the crane kind, but feeble, and united by membranes, as in those of the goose. Of what use these membranes are does not appear, as the bird is never seen swimming, its legs and thighs being of sufficient length for wading into those depths where it seeks for prey.

This bird was formerly found in great plenty on all the coasts of Europe, but it is now seen only in the retired parts of America. Its beauty, size, and the peculiar delicacy of its flesh, have been such temptations to take or destroy it, that it has long since deserted the shores frequented by man, and taken refuge in countries that are

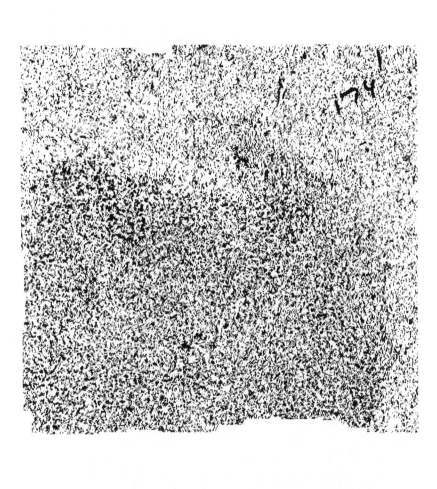

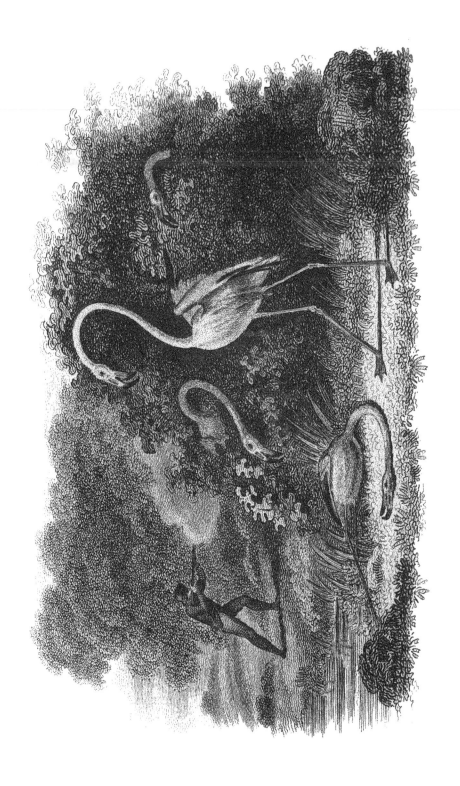

as yet but thinly peopled. In these solitary regions, the flamingo lives in a state of society, and
seemingly under a better polity than any other
of the feathered creation, as has frequently been
observed by those who have traversed that extensive continent; and who, in those extensive
journies have taken repeated notice of the good
order preserved in their retreats at the approach
of man, whom they then consider as an enemy,
by invading their territories; and in which they
live in peace and security.

Mr. Albin tells us, that when the Europeans
who first went to America, coasted down along the
African coasts, they found the flamingos on several shores on both continents, gentle, and no
ways distrustful of mankind. They had long
been used to security, in the extensive solitudes
they bad chosen; and knew no enemies but those
they could very well evade or oppose. The negroes and the native Americans, were possessed
but of few destructive arts for killing them at a
distance; and when the bird perceived the arrow,
it well knew how to avoid it. But it was otherwise when the Europeans first came among them;
the sailors, not considering that the dread of fire-
arms was totally unknown in that part of the
world, gave the flamingo the character of a foolish bird, that suffered itself to be approached
and shot at. When the fowler had killed one,
the rest of the flock, far from attempting to fly,
only regarded the fall of their companion in a

kind of fixed astonishment: another and another shot was discharged; and thus the fowler often levelled almost the whole flock, before one of them began to think of escaping. Experience, however, taught them better; at present it is very different, for the flamingo is not only one of the scarcest, but of the shyest birds in the world, and the most difficult of approach. *See the engraving.*

They chiefly keep near the most deserted and inhospitable shores; near salt-water lakes, and swampy islands. They come down to the banks of rivers by day; and often retire to the inland, mountainous parts of the country at the approach of night. When seen by mariners in the day, they always appear drawn up in a long close line of two or three hundred together, like a regiment of soldiers; or, as Dampier says, they present, at the distance of half a mile, the exact representation of a long brick wall. Their rank, however, is broken when they seek for food; but they always appoint one of the number as a watch, whose only employment is to observe and give notice of danger, while the rest are feeding. As soon as this trusty sentinel perceives the remotest appearance of danger, he gives a loud scream, with a voice as shrill as a trumpet, and instantly the whole cohort are upon the wing. They feed in silence; but, upon this occasion, all the flock are in one chorus, and fill the air with intolerable screams. Their food consists

principally· of small fish, and water insects.
These they take by plunging the bill and part of
the head into water, and from time to time tram-
pling the bottom with their feet to disturb the
mud, in order to raise up their prey. In feeding, they twist their neck in such a manner, that
the upper part of the bill is applied to the ground;
but of this we shall treat hereafter.

Notwithstanding Dampier, and others, assert,
that the flamingo at present avoids the human
race with the most cautious timidity, it is certainly not from any antipathy to man that they
shun his society, for in some villages, as Labat
declares, along the coast of Africa, the flamingos
come in great numbers to make their residence
among the natives. There they assemble by
thousands, perched on the trees, within and about
the villages; and are so very clamorous, that the
sound is heard at near a mile distance. The negroes are fond of their company; and consider
their society as a gift of heaven, and a protection
from accidental evils. They feed, protect, and
endeavour to render them every possible assistance. " But my countrymen," says M. de Buffon, " who are admitted to this part of the coast,
cannot, without some degree of discontent, see
such quantities of game, and that of the most
delicate kind, remain untouched, and rendered
useless by the superstition of the natives; they
therefore take every opportunity of killing them,
when they find themselves unobserved, and at a

convenient distance from the villages; but they
find it very necessary to hide them in the long
grass, if they perceive any of the negroes approaching; for they would probably stand a
chance of being ill treated, if the blacks discovered their sacred birds were thus unmercifully
treated, and destroyed for the purpose of indulg-
ing the appetites of their visitors."

Before they became so exceedingly shy, and
even now in some part of Africa, they are frequently shot by the mariners on the coasts, who
also not unoften are enabled to catch their young,
although they run exceedingly fast. Labat says,
that he has frequently taken them with nets, properly extended round the places they breed in.
When their long legs are entangled in the meshes,
they are then unqualified to make their escape:
but they still continue to combat with their destroyer, and the old ones, though seized by the
head, will scratch with their claws; and with
which, though to every appearance, inoffensive
weapons, they very often wound their enemies;
and even when they are fairly disengaged from
the net, they preserve a natural obstinacy and
ferocity; they refuse all nourishment; and peck
and combat with their claws at every opportunity
at those who come near them; " therefore," continues this author, " there is an absolute necessity
for destroying them, when taken, as they would
only pine and die, if an attempt was made to
keep them in captivity." The flesh of the old

Esteemed by some as delicate food.

ones is black and hard; though Dampier says,
well tasted; but that of the young ones is much
better, and esteemed as excellent by many. But
of all other delicacies, the flamingo's tongue is
the most celebrated. "A dish of flamingos'
tongues," says M. de Buffon, " is a feast for an
emperor." In fact, the Roman emperors considered them as the highest luxury; and we have
an account of one of them, who procured fifteen
hundred flamingos' tongues to be served up in a
single dish. The tongue of the flamingo, which
is so much sought after, is a good deal larger
than that of any' other bird whatever. Its bill is
like a' large black box, of an irregular figure,
and · filled with a tongue which is black and
gristly, and which has long been reckoned
among the epicures, as, a most rare ·delicacy,
from possessing a very pleasing and .peculiar flavour. Be this as it · may, a respectable author
says, " It is probable that the beauty and scarcity of the bird, might be the first inducements
to studious gluttony to fix upon its tongue as
meat for the table." What Dampier asserts of
the goodness of its flesh, cannot so well be, relied
on; for Dampier was often in want of provisions,
and then naturally thought any thing good that
could be eaten, and possibly might estimate the
delicacy of any fresh food in proportion to the
wants it happened to supply; but even he, however, agrees with Labat, that the flesh is black,
tough, and fishy; so that we can hardly give him

z 2

credit, when he asserts, that the flesh of the fla-.
luxurious entertainment.

The flamingos, as. already observed, always go
in flocks together; and when they change their
situations, they do it in. ranks in the same manner as the cranes. They are sometimes seen, at
the break of day, flying down in great numbers
from the mountains; and conducting each other
with a kind of trumpet cry, that sounds like the
word " tococo," from whence the savages of Canada gave them that name. In their flight they
appear to great advantage; for they then seem to
be of as bright a red as a burning coal. When
they dispose themselves to feed, they cease their
cry; and then they disperse over a whole marsh,
in silence and assiduity, Their manner of feeding is very singular, the bird thrusts down its
head, so that the upper convex side of the bill
shall only touch the ground; and in this position
the animal appears, as it were, standing upon its
head.

In this manner it paddles and moves the bill
about, and seizes whatever fish or insect happens
to offer. For this purpose the upper chap is
notched at the edges, so as to hold its prey with
the greater security. Catesby, however, gives a
different account of their feeding; he says that
they thus place the upper chap undermost, and
so work about, in order to pick up a seed, from
the bottom of the water, that resembles millet:

2

but as in picking up this, they necessarily also suck in a great quantity of mud; their bill is toothed at the edges, in such a manner as to let out the mud, while they swallow the grain.

Their time of breeding is according to the climate in which they reside; in North America they breed in our summer; on the other side the line they take the most favourable season of the year. They build their nests in extensive marshes, and where they are in no danger of a surprise. The nest is not less curious than the animal that builds it: it is raised from the surface of the pool about a foot and a half, formed of mud, scraped up together, and hardened by the sun, or the heat of the bird's body; it resembles a truncated cone, or that of a hillock, with a cavity at the top; the hillock being of such a height as to admit of the bird's sitting on it, or rather standing, as her legs are placed one on each side at full length: on the top it is hollowed out to the shape of the bird, and in that cavity the female lays eggs, without any lining but the well cemented mud that forms the sides of the building. She always lays two eggs, and no more, which are white. Linnæus says, that she will sometimes lay her eggs on a projecting part of a low rock, if it happen to be sufficiently convenient to admit of the legs being placed in this manner on each side.

The young ones are a long while before they are able to fly; but they very soon run with amazing swiftness. They are sometimes caught; and, very different from the old ones, suffer themselves to be carried home, and are tamed very easily. In five or six days they become so familiar as to eat out of the hand; they drink a surprising quantity of sea-water, and of which it is necessary to give them plenty. But though they are easily rendered domestic they are not reared without the greatest difficulty; for they generally pine away, for want of their natural supplies, and mostly die in a short time. While they are yet young, their colours are very different from those lively tints they acquire with age. In their first year they are covered with plumage of a white colour, mixed with grey; in the second year the whole body is white, with here and there a slight tint of scarlet; and the great covert feathers of the wings are black; the third year the bird acquires all its beauty; the plumage of the whole body is scarlet, except some of the feathers in the wings, that still retain their sable hue. Of these beautiful plumes, the savages make various ornaments; and they were formerly transported into Europe for the purpose of making muffs, but are at present almost in disuse, and preserved only as ornaments by the curious.

These beautiful birds were much esteemed by the Romans, who often used them in their grand sacrifices, and sumptuous entertainments.

CHAP. VII.

" . . . . . . The *swan,* with arched neck
Between her white wings mantling, proudly rows
Her state with oary feet."

MILTON.

THE SWAN.

SO much difference is there between this bird
when on land and in the water, that it is hardly
to be supposed the same, for in the latter no one
can possibly exceed it for beauty and grandeur.
When it ascends from its favourite element, its
motions are aukward, and its neck is stretched
forward with an air of stupidity; but when seen
smoothly sailing along the water, commanding a
thousand graceful attitudes, and moving at pleasure without the smallest effort, there is not a
more beautiful figure in all nature. In the exhi-
bition of its form, there are no broken or harsh
lines; no constrained or catching motions; but
the soundest contours, and the easiest transitions;
the eye wanders over every part with insatiable

pleasure, and every part takes a new grace with new motion. It will swim faster than a man can walk.

This bird has long been rendered domestic; and it is now a doubt whether there be any of the tame kind in a state of nature. The colour of the tame swan is entirely white, and it generally weighs full twenty pounds. The windpipe sinks down into the lungs in the ordinary man-

tribes; it can do nothing more than hiss, which it does on receiving any provocation. In these respects it is very different from the wild or whistling swan.

This beautiful bird is as delicate in its appetites, as elegant in its form. Its chief food is corn, bread, herbs growing in the water, and roots and seeds, which are found near the margin. At the time of incubation it prepares a nest in some retired part of the bank, and chiefly where there is an islet in the stream. This is composed of water-plants, long grass, and sticks;

great assiduity. The swan lays seven or eight eggs, white, one per day, much larger than those of a goose, with a hard, and sometimes a tuberous shell. It sits near two months before its young are excluded; which are ash-coloured when they first leave the shell, and for some months after. It is not a little dangerous to approach the old ones, when their little family are feeding round them. Their fears, as well as their pride, seem to take the alarm, and when in danger, the old birds carry off the young ones on their back.

Dr. Latham says, that he knows two females, that for three or four years past have agreed to associate; and have each a brood yearly, bringing up together about eleven young; they sit by turns, and never quarrel. When a twelve-month old the young swans change their colour with their plumage. All the stages of this bird's approach to maturity are slow, and seem to mark its longevity. It is two months hatching; a year in growing to its proper size; and if, according to the observations of Pliny, Buffon, and other naturalists, that those animals which are longest in the womb are the longest lived, the swan must exceed in length of years every other, for it is the longest in the shell of any bird hitherto known, and indeed has been long remarkable for its longevity. A goose, Mr. Willoughby observes, has been known to live an hundred years; and the swan, from its superior size, and from its harder, and firmer flesh, may naturally be supposed to live still longer. It is a very strong bird, and at times extremely fierce: it has not unfrequently been known to throw down and trample upon youths of fifteen or sixteen years of age; and an old swan, we are told, is able to break the leg of a man with a single stroke of its wing. A female, while in the act of sitting, observed a fox swimming towards her from the opposite shore: she instantly darted into the water, and, having kept him at bay for a considerable time with her wings, at last succeeded in drowning him; after which, in the sight of several persons, she returned in triumph. This circumstance took place at Pensy, in Buckinghamshire; and is illustrated by an engraving, which accompanies this article.

Swans were formerly held in such great esteem in England, that, by an act of Edward the Fourth, none, except the son of the king, was permitted to keep a swan, unless possessed of five marks a year. By a subsequent act, the punishment for taking their eggs was imprisonment for a year and a day, and a fine at the king's will. At present they are but little valued for the delicacy of their flesh; but many are still preserved for their beauty. Multitudes may be seen on the Thames, where they are esteemed royal property, and it is accounted felony to steal their eggs.

The flesh of the old bird is hard, and ill tasted; but those of the young, or cygnets, are still fattened near Norwich, chiefly for the tables of the corporation of that place. Persons who have property on the river there, take the young birds, and send them to some one who is employed by the corporation, to be fed; and for his trouble he is paid about half a guinea per bird. They were, a few years ago, valued at a

Description.

guinea a piece; but when sold, they now bring much more.

At Abbotsbury, in Dorsetshire, there was formerly a noble swannery, the property of the earl of Ilchester, where six or seven hundred birds were kept; but, from the mansion being almost deserted by the family, this collection has of late years been much diminished. The royalty belonged anciently to the abbot, and previous to the dissolution of the monasteries, there were frequently above double this number.

WILD, OR WHISTLING SWAN.

THIS is somewhat smaller than the tame swan, and in weight seldom exceeds sixteen pounds. The bill is three inches long; yellowish white to the middle, but black at the end. The whole plumage is white, but along the back, and the tips of the wings, are ash-colour; and the legs are black.

The windpipe, after a strange and wonderful contortion, enters through a hole, formed in the breast-bone; and being reflected therein, returns by the same aperture; and being contracted into a narrow compass by a broad and bony cartilage, it is divided into two branches, which, before they enter the lungs, are dilated, and as it were, swollen out into two cavities. By this curious construction, this bird is enabled to utter a loud and shrill note.

This species is an inhabitant of the northern regions; never appearing in England except in hard winters, when flocks of five or six are now and then seen. Martin says, that in the month of October, Swans come in great numbers to Lingey, one of the Western Isles, and continue there till March, when they return northward to breed. A few continue in Mainland, one of the Orkneys, and breed in the little islands of the fresh water lochs; but the principal part of them retire at the approach of spring. They are called the countryman's almanack; for their quitting

their arrival the reverse.

In Iceland, these birds are an object of chase. In the month of August they lose their feathers to such a degree as not to be able to fly. The natives, at that season, resort in great numbers to the places where they most abound; and are accompanied with dogs, and active strong horses, trained to the sport, and capable of passing nimbly over the boggy soil and marshes. The swans will run as fast as a tolerable fleet horse. The greater number are taken by the dogs; which are taught to seize them by the neck; a mode of attack that causes them to lose their balance, and become an easy prey.

Notwithstanding their size, these birds are so

extremely swift on the wing, when in full feather, as to make them more difficult to be shot than almost any other; it being frequently necessary to aim ten or twelve feet before their bills. This, however, is only when they are flying before the wind in a brisk gale; at which time they seldom proceed at the rate of less than a hundred miles an hour; but when flying across the wind or against it, they are not able to make any great progress.

This species has several distinctions from that of the tame swan. "Such," says Buffon, "is the extraordinary difference between these two animals, which seem to be of one species. Whether it is in the power of long continued captivity and domestication to produce this strange variety between birds, otherwise the same, I will not take upon me to determine. But certain it is, that our tame swan is no where to be found, at least in Europe, m a state of nature."

The whistling swan emits its loud notes only when flying, or calling: its sound is, " whoogh, whoogh," very loud and shrill, but not disagreeable when heard high in the air and modulated by the winds. The Icelanders compare it to the notes of the violin: they hear it at the end of their long and gloomy winter, when the return of the swans announces also the return of summer; every note therefore must be melodious which presages a speedy thaw, and a release from their tedious confinement.

It was from this species alone that the ancients derived their fable of the swan's being endued with the powers of melody. Embracing the Pythagorean doctrine, they made the body of this bird the mansion of the souls of departed poets; and then attributed to the birds the same faculty of harmony which they had thus possessed in a not distinguishing between sweetness of numbers and melody of voice, thought that real which was only intended figuratively. The mute or tame swan never frequents the Padus; " and I am almost equally certain," says Mr. Pennant, " that it never was seen on the Cayster, in Lydia; each of them; streams celebrated by the poets for the great resort of swans. The Padus was styled oloriferous, from the numbers of these birds which frequent its waters; and there are few of the poets, either Greek or Latin, who do not make them its inhabitants."

THE DUCK.

THE common duck, of which there are about ten different sorts, is so universally known as to require no description. It is the most easily reared of all our domestic animals. The very instinct of the young ones direct them to their favourite element; and though they are conducted by a hen, they despise the admonition of their leader. The feet of the tame duck are black.

It is usual to lay duck eggs under a hen, because she hatches them better than the original parent would have done. The duck seems to be, a heedless, inattentive mother; she frequently leaves her eggs till they spoil, and even seems to forget that she is entrusted with the charge: she is equally regardless of them when excluded; she leads them to the pond, and thinks she has sufficiently provided for her offspring when she has shown them the water. Whatever advantages may be procured by coming near the house, or attending in the yard, she declines them all; and often lets the vermin, who haunt the waters destroy them, rather than take shelter nearer home. The hen is a nurse of a very opposite character; she broods with the utmost assiduity, and generally brings forth a young one from every egg committed to her charge; she does not lead them to the water indeed, but she carefully guards them when there by standing at the brink. Should the rat or the weazel attempt to seize them, the hen instantly gives them protection; she leads them to the house, when tired with paddling, and rears up the suppositious brood, without ever suspecting that they belong to another.

THE EIDER DUCK.

THIS species is about twice the size of the common duck. Its bill is black and cylindrical; the feathers of the forehead and cheeks advance far into the base. In the male, the feathers of part of the head, of the lower part of the breast, the belly, and the tail, are black, as are also the quill-feathers of the wings; and nearly all the rest of the body is white. The legs are green. The female is of a reddish brown, variously marked with black and dusky streaks. It is principally found in the western isles of Scotland, on the coasts of Norway, Iceland, and Greenland, and in many parts of North America, particularly in the Esquimaux Islands.

In Iceland, the eider ducks generally build their nests on small islands not far from the shore; and sometimes even near the dwellings of the natives, who treat them with so much attention and kindness, as to render them nearly tame. Sometimes two females will lay their eggs in the same nest, in which case they always agree remarkably well. The female lays from three to five eggs (sometimes so many as eight) which are large, smooth, glossy, and of a pale olive colour. They generally lay among stones, or plants, near the sea, but in a soft bed of down, which they pluck from their own breasts.

As long as the female is sitting, the male con-

3

Mode of procuring the down.

tinues on watch near the shore; but as soon as
the young are hatched, he leaves them. The
mother, however, remains with them a considerable time afterwards. It is curious to observe
her manner of leading them out of the nest, almost as soon as they creep from the eggs.
Going before them to the shore, they trip after
her: and, when she comes to the water-side, she
takes them on her back, and swims a few yards
with them, when she dives; and the young ones
are left floating on the surface, obliged to take
care of themselves. They are seldom seen afterwards on land.

From these birds is produced the soft down so
well known by the name of eider, or edder down,
with which (as before observed) they line the inside of their nests, which renders them particularly
warm. When the natives come to the nest, they
carefully remove the female, and take away the su-
perfluous down and eggs; after this they replace
the female: she then begins to lay afresh, and
covers her eggs with new down, which she also
plucks from her body; when this is scarce, or
she has no more left, the male comes to her assistance, and covers the eggs with his down,
which is white, and easily distinguished from that
of the female. When the young ones leave the
nest, which is about an hour after they are
hatched, it is once more plundered.

The most eggs, and the best down, are got
during the first three weeks of their laying; and

it has generally been observed, that they lay the greatest number of eggs in rainy weather. One female, during the time of laying; generally gives half a pound of down; which, however, is reduced one-half after it is cleansed.

The cider-down is of such value, when in its purity, that it is sold in Lapland for two rix-dollars a pound. It is extremely soft and warm; and so light and expansive, that a couple of handfuls squeezed together are sufficient to fill a down quilt, which is a covering like a feather-bed, used in those cold countries instead of a common quilt or blanket.

Fifteen hundred, or two thousand pounds weight of down, cleansed and uncleansed, are generally exported from Iceland, every year, by the Iceland company at Copenhagen, exclusive

the year 1750, this company sold so much in quantity of this article, as produced 3747 rix-dollars, besides what was sent directly to Gluckstadt.

The Greenlanders kill these birds with darts; pursuing them in their little boats, watching their course by the air bubbles when they dive, and always striking at them when they rise to the surface wearied. The flesh is valued as food, and their skins are made into warm and comfortable under-garments.

THE WILD DUCK.

THERE are about twenty different sorts of the wild duck, according to Buffon, and they differ from the tame by hairy yellow feet.

Wild ducks frequent the marshy places in many parts of this kingdom; but no where in greater plenty than in Lincolnshire.

Numerous as the varieties of wild ducks may be, they all pursue the same mode, and live in the same manner, keeping together in flocks in the winter, and flying in pairs in summer, bringing up their young by the water-side, and leading them to their food as soon as out of the shell. The nest, whether high or low, is generally composed of singular materials. The longest grass, mixed with heath, and lined within with the bird's own feathers, usually go to the composition; however, in proportion as the climate is colder, the nest is more artificially made, and more warmly lined. In the Arctic regions, nothing can exceed the great care all of this kind take to protect their eggs from the intenseness of the weather. While the gull and the penguin kind seem to disregard the severest cold, the duck, in those regions, forms itself a hole to lay in, shelters the approach, lines it with a layer of long grass and clay, within that another of moss, and lastly, a warm coat of feathers or down.

As the whole of this tribe possess the faculties

of flying and swimming, so they are in general birds of passage, and it is most probable, perform their journies across the ocean as well on the water as in the air. Those that migrate to this country, on the approach of winter, are seldom found so well tasted, or so fat, as the fowls that continue with us the year round: their flesh is often lean, and still oftener fishy; which flavour it has probably contracted in the journey, as their food in the lakes of Lapland, from whence they descend, is generally of the insect kind.

As soon as they arrive in this country, they are generally seen flying in flocks to make a survey of those lakes where they intend to take up their residence for the winter. In the choice of these they have two objects in view; to be near their food, and yet remote from interruption. Their chief aim is to choose some lake in the neighbourhood of a marsh, where there is at the same time a cover of woods, and where insects are found in great abundance. Lakes, therefore, with a marsh on one side, and a wood on the other, are seldom without vast quantities of wild fowl; and when a couple are seen at any time, that is a sufficient inducement to bring hundreds of others. The ducks, flying in the air, are often lured down from their heights by the loud voice of the mallard below. Nature seems to have furnished this bird with very particular faculties for calling. The windpipe, where it begins to enter the lungs, opens in a kind of bony cavity,

where the sound is reflected as in a musical instrument, and is heard a great way off. To this
call all the stragglers resort; and in a week or a
fortnight's time, a lake that was before quite
naked, is black with water-fowl, that have left
their Lapland retreats to keep company with our
ducks who never stirred from home.

"They generally," observes a celebrated author, "choose that part of the lake where they
are inaccessible to the approach of the fowler,
in which they all appear huddled together, extremely busy and very loud. What it is can
employ them all the day is not easy to guess.
There is no food for them at the place where
they sit and cabal thus, as they choose the middle of the lake; and as for courtship, the season
for that is not yet come; so that it is wonderful
what can so busily keep them occupied. Not
one of them seems a moment at rest. Now pursuing one another, now screaming, then all up
at once, then down again; the whole seems one
strange scene of bustle with nothing to do.

"They frequently go off in a more private
manner by night to feed in the adjacent meadows and ditches, which they dare not venture
to approach by day. In these nocturnal adven-
tures they are often taken; for though a timorous bird, yet they are easily deceived, and every
spring seems to succeed in taking them. But
the greatest quantities are taken in decoys;
which, though well known near London are yet

untried in the remoter parts of the country." In only ten decoys in the neighbourhood of Wainfleet, as many as thirty-one thousand two hundred have been caught in one season. Numbers are annually taken thus in Lincolnshire.

A decoy is a pond generally situated in a marsh, so as to be surrounded with wood or reeds, and if possible with both, to prevent the birds which frequent it from being disturbed. In this pond the birds sleep during the day; and as soon as the evening sets in, the decoy rises (as it is termed), and the wild fowl feed during the night. If the evening be still, the noise of their wings during flight is heard at a great distance, and is a pleasing though somewhat melancholy sound. The decoy-ducks (which are either bred in the pond-yard, or in the marshes adjacent; and which, although they fly abroad, regularly return for food to the pond, and mix with the tame ones that never quit the pond) are fed with hemp-seed, oats, and buck-wheat. In catching the wild birds, hemp-seed is thrown over the skreens to allure them forward into the pipes; of which there are several, leading up a narrow ditch, that closes at last with a funnel-net. Over these pipes, which grow narrower from the first entrance, there is a continued arch of netting, suspended on hoops. It is necessary to have a pipe for almost every wind that can blow, as on that circumstance it depends which pipe the fowl will take to. The decoy-man likewise always

keeps to the leeward of the wild fowl: and burns
in his mouth, or hand, a piece of Dutch turf, that
his effluvia may not reach them ; for if they
once discover by the smell that a man is near,
they all instantly take flight. Along each pipe
are placed red skreens, at certain intervals, to
prevent him from being seen till he thinks proper
to show himself, or the birds are passed up the
pipe, to which they are led by the trained ducks
(who know the decoy-man's whistle), or are en-
ticed by the hemp-seed. A dog is sometimes,
used; who is taught to play backwards and for-
wards between the skreens, at the direction of his
master. The fowl, roused by this new object,
advance towards it, while the dog is playing still
nearer to the entrance of the pipes; till at last
the decoy-man appears from behind the skreens,
and the wild-fowl, not daring to pass by him,
and unable to fly off on account of the net covering the hoops, press forward to the end of the
funnel-net which terminates upon the land, where
a person is stationed ready to take them. The
trained birds return back past the decoy-man,
into the pond again, till a repetition of their ser-
vices is required. The general season for catching, is from the latter end of October till February. There is a prohibition, by act of parliament, against taking them between the first of
June and the first of October.

It was formerly customary to have, in the fens,
an annual driving of the young ducks, before

2

they took wing.   Numbers of people assembled,
who beat a vast tract, and forced the birds into a
net, placed at the spot where the sport was to ter-
minate.   By this practice (which, however, has
been abolished by parliament,) as many as a hundred and seventy-four dozen have been known to
be taken in one day.

Prodigious numbers of these birds are taken by
decoys in Picardy in France, particularly on the
river Somme.   It is customary there, to wait for
the flock's passing over certain known places;
when the sportsman, having a wicker cage containing a quantity of tame birds, lets out one at
a time, which enticing the passengers within gun-
shot, five or six are often killed at once, by an
expert marksman.   They are now and then also
taken by hooks, baited with raw meat, which the
birds swallow while swimming on the water.

Other methods of catching ducks and geese
are peculiar to certain nations: one of these,
from its singularity, seems worth mentioning.   A
person wades into the water up to the chin; and
having his head covered with an empty calabash,
approaches the place where the ducks are;
which, not regarding an object of this kind, suffer the man freely to mix with the flock; when
he has only to pull them by the legs under the
water, one after another, and fix them to his belt,
till he is satisfied: returning as unsuspected by
the remainder as when he first came among
them.   This curious method is frequently prac-

tised on the river Ganges, the earthern vessels of
the Gentoos being used instead of calabashes.
These vessels are what the Gentoos boil their
rice in: after having been once used, they are
looked upon as defiled, and are thrown into the
river as useless: the duck-takers find them convenient for their purpose; as the ducks, from
seeing them constantly float down the stream,
look upon them as objects not to be regarded.

Wild ducks are very artful birds. They do
not always build their nest close to the water;
but often at a good distance from it; in which
case the female will take the young in her beak,
or between the legs, to the water. They have
been known sometimes to lay their eggs in a high
tree, in a deserted magpie or crow's nest; and an
instance has likewise been recorded of one being
found at Etchingham, in Sussex, sitting upon
nine eggs, in an oak, at the height of twenty-five
feet from the ground: the eggs were supported
by some small twigs, laid crossways.

At Bold, in Lancashire, it is said there were
formerly great quantities of wild ducks, during
the summer time, in the ponds and moat near
the hall; which used regularly to be fed. A man
beat with a stone on a hollow wooden vessel, and
immediately the ducks would come round him.
He scattered corn among them, which they gathered with as much quietness and familiarity as
might be expected from tame ducks. As soon

as they had finished their repast, they returned to their accustomed haunts.

The Chinese make great use of ducks, but prefer the tame to the wild ones. It is said that the major part of the ducks in China are hatched by artificial heat. The eggs, being laid in boxes of sand, are placed on a brick hearth, to which is given a proper heat during the time required for hatching. The ducklings are fed with crawfish and crabs, boiled and cut small, and afterwards mixed with boiled rice ; and in about a fortnight they are able to shift for themselves. The Chinese then provide them an old step-mother, who leads them where they are .to find provender ; being first put on board a sampane, or boat, which is destined for their habitation ; and from which the whole flock, often to the amount of three or four hundred, go out to feed, and return at command. This method is used nine months out of the twelve, (for in the colder months it does not succeed ;) and is so far from a novelty, that it may be every where seen : but more especially about the time of cutting the rice, and gleaning the crop ; when the masters of the duck-sampanes row up and down the river, according to the opportunity of procuring food, which is found in plenty, at the tide of ebb, on the rice plantations, as they are overflowed at high water. It is curious to observe how the ducks obey their masters ; for some thousands,

belonging to different boats, will feed at large on
the same spot, and on a signal given, will follow
their leader to their respective sampanes, without
a single stranger being found among them. This
is still more extraordinary, if we consider the
number of inhabited sampanes on the Tigris;
supposed to be no less than forty thousand,
which are moored in rows close to each other,
with here and there a narrow passage for boats
to sail up and down the river. The inhabited
sampanes contain each a separate family, of
which they are the only dwelling; and many of
the Chinese pass almost their whole lives in this
manner on the water. The Tigris at Canton is
somewhat wider than the Thames at London,
and the whole river is there covered in this manner for the extent of at least a mile.

THE COMMON WILD DUCK, OR BOCHAS.

FROM this species the tame ducks take their
origin, and to which they may be traced by unerring characters. The intermediate tail feathers
of the drake are turned backwards, and the bill
is straight, two circumstances that universally
prevail in the same sort. The difference of taste
is easily accounted for, from the difference of
their food. They pair in the spring, build their
nests among the rushes near the water, and lay

from ten to sixteen eggs. The female is a very artful bird, especially where the safety of her young are at stake. In summer they fly in pairs, bring up their young by the water-side, and lead them to food as soon as they are out of the shell. When apprehensive of danger, they have been known to build their nests in high trees, and in other birds' nests. A tmoulting time, when they cannot fly, they are caught in great plenty; and in their annual migration to this country, they are taken in decoys in still greater abundance, particularly in Lincolnshire, the grand magazine of wild fowl in this kingdom.

THE SCAUP DUCK, OR MACULA;

SO called from feeding on broken shell-fish, is less than the common duck; it is a beautiful bird, but so diversified in colouring, that scarcely two in a hundred can be found alike.

THE SHELDRAKE.

THIS has a flat bill, a compressed forehead ariega with white. It is an inhabitant of the northern world, so far as Iceland. They usually breed in deserted rabbit-holes, and lay fifteen or sixteen roundish white eggs, and sit about thirty days. " They are very careful of their young," says

Latham, " and will carry them from place to place in their bills." They also show much instinctive cunning in preserving them when attempted to be caught; for they will fly along the ground as if wounded, till the brood are got into a place of security. Their great beauty has induced many unsuccessful attempts to domesticate them; but they never thrive unless in the neighbourhood of salt water. The eggs are thought good, but the flesh of this bird is rank and unsavory.

THE KING DUCK, OR GREY-HEADED DUCK,

A VERY beautiful species, is found at Hudson's Bay; it is also common in Siberia, and in Greenland, where the flesh is accounted excellent; and of the skins sewed together, the natives make very comfortable garments; nor is its down less comfortable than that of the eider.

THE SCOTER, OR NIGRA.

THE male of this species is totally black; the female brownish; the tail resembles a wedge; in the winter season they are found on the coasts of Great Britain; but are very numerous on the shores of France from November to March, where they feed upon a glossy bivalve shell, called vaimeaux, and are caught by placing nets under the water where these shells abound; to obtain which, the birds dive to a great depth, and thus thirty or forty dozen are often taken in a tide. They swallow the shells whole, which have been found quite crumbled to powder among their excrements. They are sometimes kept tame, and feed upon soaked bread. Their flesh is far from being agreeable, and is of so very fishy a taste that, perhaps by way of mortification, it is allowed to be eaten by Roman Catholics on fast days. This species is also to be met with in North America; and it abounds in the northern parts of Europe, especially on the great lakes and rivers of Siberia.

THE HOOK-BILLED DRAKE

wards, and is about two feet from the extremity of the bill to the end of the tail, and in breadth from the extension of each wing, near three feet. The bill is crooked, of a palish green, except the hook at the end, which is black; it is in length

neck, and the head, are of a dark green, with two small white speckled lines, one of which runs from the upper part of the bill, over the eye towards the back part of the head; the other runs from the bill to the lower part of the eye, around which there is a circle of fine white feathers;

with small white feathers under the chin. The
breast, belly, and throat, are white, with small
transverse spots, of a brownish red, running
across them. The six first of the prime feathers
of the wings are white, the rest of a reddish
brown; the first row of covert feathers are blue
tipped with white, the second are brown with
white tips. The scapular feathers of the wings,
the sides, and the back, are of a reddish brown,
which appears dusted or speckled over with
white. The tail is black, with white tips, which
turn up in a sort of circular curl towards the
back. The legs and feet are of a fine orange
colour.

THE MALLARD

IS about the size of the preceding; its bill,
from the angles of the mouth to the tip, is about
two inches and a quarter, and near an inch broad,
with a roundish tip at the end; the head and
upper part of the neck are of a beautiful shining
green; the under eye-lids white, with a sort of
half circle, or white ring, that passes round the
fore part of the neck; the under part of the neck
below the white ring to the breast, is of a glossy
chesnut colour. The under part of the breast
and belly are a sort of ash-colour, sprinkled with
a variety of dark specks, resembling drops; the
back between the wings is of a cinerous red, in

like manner sprinkled or speckled; the lower part towards the rump still darker; the rump itself of a sort of glossy purple. The sides of the body, and the longer thigh-feathers, are beautified with transverse brown lines, with a bluish sort of mixture. The scapular feathers of the wings are of a fine silver colour, beautifully variegated with brown transverse lines; the second row of the quill-feathers tipped with white, with the outward webs of a fine bluish purple, and a border of black, running between the white and the blue; the rest of the wings variegated with silver-coloured feathers, with some of their edges black, others of a dark purple. The under parts of the tail is black, the feathers on the upper part end in sharp points, the middlemost of which turn up in a circular form towards the back, and appear of a fine glossy purple colour. They are feathered down to the knees, the legs and feet are of a saffron colour.

THE TUFTED, OR BLACK-CRESTED DUCK.

THIS is not quite so large as the wild mallard, the shape of its body appearing more broad, short, thick, and compressed; the bill broad, and about two inches long, is of a palish blue colour, black at the tip; the upper part of the head is of a blackish mixed purple, with a fine crest of feathers hanging down behind the head, of near two inches long; the nostrils are pretty large, and the irides of the eyes of a gold colour, or fine yellow. The neck and upper part of the body are of a dark brown, much inclining to black. The wings are short with black covert feathers, the outward wings of the same colour, by degrees growing more towards a white; the second row of quills is all white, with black tips. The under parts of the neck and the breast are black, the belly of a fine silver-coloured white, as are also the thighs, and under parts of the wings. The tail is short, composed of black feathers; the legs are short, and the feet of a dark lead colour.

Mr. Albin says, that at Venice, and other parts of Italy, this bird goes by the name of Cape Negro.

THE UPRIGHT DUCK:

SO called, as it walks in a more stately and erect posture than any other of this kind. Its bill is of a greenish colour, with a sort of brown shade or cast; the circles of the eyes are white; the top of the head is quite black, under which, from the upper base of the bill, there runs a white circle which surrounds the top parts of the head; the other parts of the head are of a dark colour, intermixed with shades of red and green, which by the reflection of different lights, appear

with white and black feathers; the wing-feathers

back is of a dark colour, intermixed with beautiful shades resembling the rainbow; the sides of
the body underneath the wings, the thighs, and
near the vent, are of a sooty-coloured black; the
belly and breast white; the legs and feet are of
a sort of dusky yellow.

THE MUSCOVY DRAKE

IS considerably larger than the generality of
fowls of the duck kind; some of them being as
large as a small sized goose; the bill is broad
and short, of a reddish colour, a little hooked at
the end; upon the upper part of which, between
the nostrils, there grows a small round fleshy excrescence, that appears red like a small cherry;
the irides of the eyes are white, encircled with a
fleshy sort of red substance, resembling that on
the bill. The upper part of the head and neck
appear of a dusky colour, a little mottled with
white; the sides of the wings and the hack are
of a very uncommon mixture of red, green,
brown, purple, and white; the under part of the
body white, interspersed here and there with a
few small brown feathers; the legs and feet of a
pale red, or rather orange colour.

The flesh of the Muscovy drake differs from

that of the common duck, and is said to be much more pleasant; they lay a great many eggs, and are excellent breeders; the hen has not the tuberous flesh growing on her bill, but with respect to colour, is much the same as the cock.

The ambassador from the duke of Holstein, in his travels to Muscovy, says, he saw there a sort of wild ducks, bigger than ours, but as black as crows, with long necks, and forked bills. They are called by the Muscovites, braclan, and are scarcely ever seen but in the night time: their quills are harder and bigger than those of a crow.

THE MADAGASCAR DUCK

IS in size very little larger than the common tame duck, and has a yellowish brown bill; the circles or irides of the eyes, are red; the head and neck of a dark green; the breast and lower parts of the body more inclining to a dusky brown; the outward edges of the feathers red; the scapular feathers are some of them green, with red edges; others more dusky, with a beautiful bluish mixture; the first row of the covert feathers are pretty much of the same colour, the second row green; the quill feathers are all beautifully edged with red; the whole mixture of the colours shine with a curious and uncommon gloss, and appear

---

---

exceedingly beautiful. The legs and feet are of
an orange colour.

They are brought from Madagascar, in the
East Indies, and are now bred by the curious in
several parts of England.

THE SHOVELLER

WEIGHS very near two pounds, and is from
the point of the bill to the end of the tail, one or
two and twenty inches, and upwards of eighteen
from the extremity of each wing when extended.
The bill is of a fine black, considerably broader
at the tip than at the base; dented in the middle,
and rising towards the end, with a small sort of
crooked hook bending downwards, each mandible being toothed like a comb; the tongue is
broad and fleshy especially towards the end, the
tip of which is of a sort of semicircular form.
The circles of the eyes are of a fine yellow; the
neck and head of a shining dark green, the crop
and under part of the neck white, the upper part
of the shoulders of the same colour, but interspersed with a variety of bold strokes; the under
part of the body red, except the feathers under
the tail, behind the vent, which are black; the
back is of a brown colour, beautifully shaded
with a shining green purple and blue, which varies according to the light in which it is viewed.

The first ten or twelve quill feathers are quite brown; the next in the same row have all their extreme edges of a shining deep green, some of them varied with small white lines; others are green, with white tips, which, when viewed together, appear like a sort of cross bar upon each wing; the covert feathers are many of them of a fine blue, others more inclining to an ash colour; the tail is composed of party-coloured feathers, some of the borders entirely white, others on their extreme edges wholly black. The thighs are interspersed with a considerable number of dusky coloured transverse lines; the legs and feet of a fine red; resembling the colour of vermillion; the claws black.

The hen bears a near resemblance to the cock in the shape of its body, but differs very much in colour; the wings are pretty much like those of the male, only the colours are more faint, and the shades not near so beautiful. The head, neck, and almost all the rest of the body, both for colour and shape, very much resemble that of the wild duck. The membrane that connects the toes of each of them are serated about the edges, and their feet seem to be considerably less than the generality of the duck kind.

THE GOLDEN FYE

CHIEFLY breeds in Italy; it has a large head and thick body; the neck short, and the bill broad, elevated towards the point, of a black colour, and is, if measured from the angles of the mouth, about an inch and three quarters long; the head, when variously exposed to the light, appears black, purple, and green, with a fine shining silky gloss; it has a white spot on each side of the mouth; the eyes are of a fine gold colour; the neck, breast, and belly, white; the space between the shoulders and the back is black; the wings of a fine beautiful mixture of black and white. The tail near three inches long; the legs short, of a yellowish colour; the toes pretty long, and more dusky.

It has a disagreeable fishy taste; they are sometimes, but very rarely, taken upon the English coast.

THE PINTAIL,

THOUGH to appearance nearly as long as the golden fye, seldom weighs more than a pound and a half; the wing-feathers are very long; the upper mandible is of a bluish black, mostly so about the nostrils, the under quite black. The neck is longer than the generality

of birds of this kind; it is slender, and of a
brown colour, very much resembling that of
rusty iron, with a tincture of purple behind
the ears; on each side of which from the hinder
parts of the head, there runs a white line, which
passes down the sides of the neck; the feathers between the white lines are black, under
which the neck is of an ash colour; both the
back and neck varied with black and white
transverse lines; the middle parts of the scapular feathers of the wings are black, their inner
parts varied with a mixture of white, black, and
brown lines, some of the tips of the second row
of feathers white; others party-coloured, with
shades of glossy red. The breast and lower parts
of the body, as far as the vent, are white; the
under part of the tail, black; the thighs more
pale, and varied with small specks of black; the
two middlemost feathers of the tail are extended
much longer than the rest, running into sharp
points, from whence it is said to take the name
of pintail; the upper part of the tail is of a sort
of ash colour, the tips of the feathers black. The
feet are of a lead colour.

THE WIDGEON

WEIGHS near a pound and a half; it has a sort of black nail at the end of the upper mandible of the bill, the other part of which is of a lead colour; the structure of the head and mouth very much resemble the common wild duck, only the head does not seem to be quite so large, in proportion to the body, which also appears of a finer shape, and the wings longer. The crown of the head towards the base of the bill, is of a pale pink colour, inclining to a reddish white; the other parts of the head and the neck, are red; the sides of the body and the upper part of the breast, are tinctured with a very fair glossy, and beautiful claret colour, with a few small transverse lines of black. The feathers on the back are brown, the edges more pale or ash-coloured; the scapular feathers, and those under the fore part of the wings, are finally variegated with small transverse black and white lines, beautifully dispersed like waves; the quill-feathers are some of them brown, with white tips, others have their outward webs of a blackish purple; other parts, especially those beyond the covert feathers, of a lovely fine blue; some of the exterior feathers have their outward webs inclining to black, with a fine purple gloss upon

small light coloured spots; the rest of the wing

feathers are of a beautiful party-coloured brown and white. The upper part of the tail is ash-coloured; the under part behind the vent, black. The legs and feet are of a dark lead colour, and the claws black.

Widgeons are common in Cambridgeshire, the Isle of Ely, &c. where the male is called the widgeon, and the female, the whewer. They feed upon wild periwinkles, grass, weeds, &c. which grow at the bottom of rivers and lakes. Their flesh has a very delicious taste, not inferior to teal, or wild ducks.

THE GREAT-HEADED WIDGEON.

THIS is larger than the common widgeon, and the make of its body is considerably thicker and shorter, weighing often near two pounds when well fed; the bill is considerably larger and broader than that of the widgeon; the head and the greatest part of the neck, are of a fine fulvous red; the feathers from the upper part of the head, come down in the form of an acute angle, or peak, to the middle of the base of the upper mandible, which is of a lead colour, tipped with black, the under mandible being entirely black; the circles, or irides of the eyes, are of a fine yellow. The small covert feathers of the wings, and those on the middle part of the back, are variegated with brown and cinerous, elegant

ıving lines. The rump, and feathers under the
ıl, are black, so that the tail, which is of a sort
a greyish colour, and about two inches long,
pears encircled with a blackish ring. The
ıddle part of the breast, and lower part of the
lly, very much resemble the colour of the back,
ly the lines and points are of a more pale co-
ır. The quill-feathers are of a dark ash-co-
ır; and it is remarkable, all the feathers on the
ıddle of the wings of this bird, are of one uni-
rm colour, without the different variations com-
only found in others of the kind. The feet are
a lead colour, and the membranes that connect
e toes more dark and blackish.

THE TEAL

IS the smallest bird of the duck kind, and
ıes not usually weigh more than twelve or four-
en ounces; it is about sixteen inches from the
ıint of the bill to the end of the tail, and from
e extremity of each wing, when extended, near
ıo feet. The bill is of a dark brown colour, the
ıad is considerably lighter, inclining to a bay,
ıth a large white stripe over each eye, bending
ıwnwards, towards the back part of the head:
ıe neck, back, and tail, are of a more dusky
ılour. The breast is of a dirty-coloured yellow,
terspersed with dusky transverse lines; the
ılly more bright, with yellowish brown spots:

Food—Description.

the quill-feathers of the wings are of a dusky brown, with white edges; the covert feathers appear of a fine shining green, with their tips white; the scapular feathers are more inclining to an ash-colour; the legs and feet are brown; the claws black.

These birds feed on water-plants, seeds and grass.

THE FRENCH TEAL

IS about the size of the former; the cock of this tribe has a broad black bill; the eyes are of a sort of hazel colour; the upper part of the head and neck are of a light brown, or bay, with a shining green line running from each eye to the back part of the head, with a black spot intervening between, and a white line passing under the eyes: the back, the lower part of the neck, and lines underneath the wings, are beautified with fine waving lines of black and white; the breast is more of a yellowish colour, spotted with black, that bears some resemblance to scales; the belly is of a dirty white, or grey.

The wings are of a brown, or dusky colour, some of them with white tips, and their outward edges black; others green, with yellowish edges; the covert feathers have some of them white tips, and the green coverts appear of a yellowish red; the whole beautifully variegated with different shades, that make a very agreeable appearance

2 E 2

he eye; the tail is sharp towards the end, and
ut three inches long; the legs and feet of a
ky pale colour.

heir flesh is of a delicate taste; it affords a
nourishment to the body, and amongst those
ts kind, is said to challenge the first place.

he Chinese teal of Edwards, and the summer
k of Catesby, are elegant species; the for-
r is a native of China, sometimes brought alive
England, but too tender to be reared in this
ntry. The other inhabits Mexico and some
the West-India islands: and is to be seen here
imes in the menageries of the curious.

CHAP. VIII.

" The noisy *geese* that gabbled o'er the pool.".
GOLDSMITH.

THE WILD GOOSE.

THIS bird has a large elevated bill, of a flesh colour, tinged with yellow; the head and neck ash-coloured; the breast and belly whitish, clouded with grey, as is also the back, and the legs of a flesh colour.

Wild geese inhabit the fens of England; and are supposed not to migrate, as they do in many countries on the continent. They breed in Lincolnshire and Cambridgeshire: they have seven or eight young; which are sometimes taken, and are easily rendered tame. At the table, these birds are deemed superior to the domestic goose.

They are often seen in flocks of from fifty to a hundred, flying at very great heights, and seldom resting by day. Their cry is frequently heard while they are imperceptible from their distance above. Whether this be their note of

2

mutual encouragement, or only the necessary consequence of respiration, seems somewhat doubtful; but they seldom exert it when they alight in their journeys. On the ground they always arrange themselves in a line, and seem to descend rather for rest than refreshment; for, having continued in this manner for an hour or two, one of them with a long loud note sounds a kind of signal to which the rest always punctually attend, and rising in a group they pursue their journey with alacrity: Their flight is conducted with vast regularity: they always proceed either in a line a-breast, or in two lines joining in an angle at the middle. In this order they generally take the lead by turns; the foremost falling back in the rear when tired, and the next in station succeeding to his duty. Their track is generally so high, that it is almost impossible to reach them from a fowling-piece; and even when this can be done, they file so equally that one discharge very seldom kills more than a single bird.

They breed in the plains and marshes about Hudson's Bay, in North America: in some years the young ones are taken in considerable numbers; and at this age they are easily tamed. It is, however, extremely singular, that they will never learn to eat corn, unless some of the old ones are taken along with them; which may be done when these are in a moulting state.

THE DOMESTIC GOOSE.

THE common tame goose is nothing more than the wild goose in a state of domestication. It is sometimes found white, though much more frequently verging to grey; and it is a dispute among men of taste, which should have the preference.

These birds, in rural economy, are an object of attention and profit, and are no where kept in such vast quantities as in the fens of Lincolnshire; several persons there having as many as a thousand breeders. They are bred for the sake of their quills and feathers; for which they are stripped while alive, once in the year for their quills, and no less than five times for the feathers: the first plucking commences about Lady-day, for both; and the other four are between Lady-day and Michaelmas. It is said that in general the birds do not suffer much from this operation; except cold weather sets in, which then kills great numbers of them. The old geese submit quietly to be plucked, but the young ones are very noisy and unruly. Mr. Pennant says he once saw this business performed, and observed, that even goslins of only six weeks old were not spared—for their tails were plucked, as he was told, to inure them early to the custom. The possessors, except in this cruel practice, treat their birds with great kindness; lodging them very often even in the same room with themselves. Should the season prove cold, vast numbers die by this savage practice.

These geese breed in general only once a-year, but if well kept they sometimes hatch twice in a season. During their sitting, each bird has a space allotted to it, in rows of wicker pens placed one above another; and it is said that the gozzard, or goose-herd, who has the care of them, drives the whole flock to water twice a day, and, bringing them back to their habitations, places every bird (without missing one) in its own nest.

It is scarcely credible what numbers of geese are driven from the distant counties to London for sale; frequently two or three thousand in a drove; and in the year 1783 one drove passed through Chelmsford, in their way from Suffolk to London, that contained about nine thousand. Among these are several superannuated geese, that in consequence of repeated pluckings, prove remarkably tough and dry.

A goose well fed in the common way, will weigh fifteen or sixteen pounds, but by the unnatural practice of cramming, may be increased to almost double that weight. The creatures set apart for this beastly and unwholesome gorge are nailed to the floor, by the webs of their feet, to keep them in a state of perfect inaction;

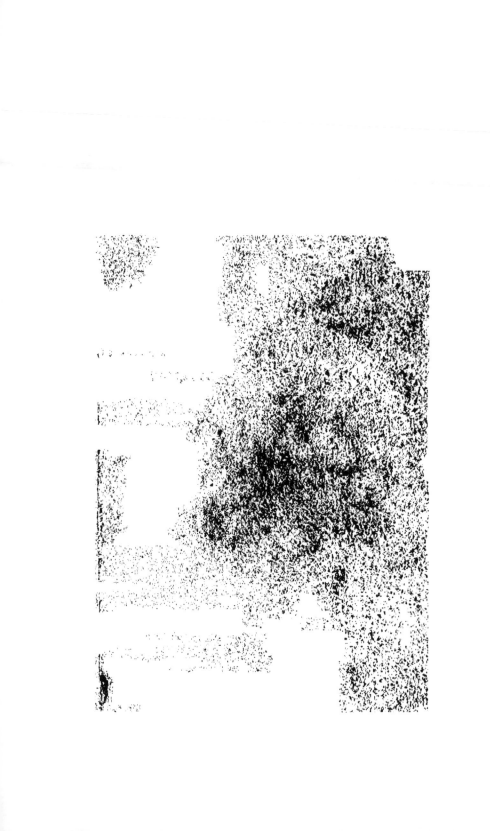

and then stuffed with bean-meal, and other fat-
tening diet; but French refinement has increased
the barbarity, by putting out the eyes of the
wretched animal.

However simple in appearance, or awkward
in gesture, the goose may be, it is not without
many marks both of sentiment and understanding. The courage with which it protects its
young and defends itself against the ravenous
birds, and certain instances of attachment and
even of gratitude which have been observed in
it, render our general contempt of the goose ill-
founded. The following instance of warm affection, which was communicated to the Comte de
Buffon by a man of veracity and information, is
here given as a proof of the goose's sagacity, accompanied by an engraving.

" There were two ganders, a grey and a white
one (the latter named *Jacquot*), with three females. The males were perpetually contending
for the company of these dames. When one or
the other prevailed, it assumed the direction of
them, and hindered its rival from approaching.
He who was the master during the night, would
not yield the next morning; and the two galants
fought so furiously, that it was necessary to be
speedy in parting them. It happened one day,
that being drawn to the bottom of the garden
by their cries, I found them with their necks en-
twined, striking their wings with rapidity and
astonishing force; the three females turned

round; as wishing to separate 'them,, but. without.

overthrown, and mal-treated, by the other./ I
parted them; happily for the white one, as he
would / otherwise have lost his life. Then the
conqueror. began screaming and gabbling, and
clapping 'his wings; and ran to join his mistres-
ses, giving each a noisy salute, to which the three

time round him. Meanwhile poor Jacquot, was
in a pitiable condition; and, retiring, sadly
vented at a distance his doleful cries. It was several days before he. recovered from his dejec-
tion; during which time I had sometimes occasion to. pass through the court where he strayed.
I saw him always thrust out from society; and
whenever I passed, he came gabbling to me.
One day he approached so near, and showed so
much friendship, that I could not help caressing
him, by stroking with my hand his back and
neck; to which he seemed so sensible, as to fol-.
low me into the entrance of the court. Next
day, as I again passed, he ran to me and I gave
him the same caresses; with which alone he was
not satisfied, but seemed, by his gestures, to desire that I should introduce him to his mates. I
accordingly led him to their quarter; and, upon
his arrival, he began his vociferations, and directly addressed the three dames, who failed not
to answer him. Immediately his late victor
sprung upon Jacquot. I left them for a mo-

ment; the grey one was always the stronger! I took part with my Jacquot, who was under; I set him over his rival; he was thrown; I set him up again. In this way they fought eleven minutes; and, by the assistance which I gave him, he at last obtained the advantage, and got possession of the three dames. When my friend Jacquot saw himself master, he would not venture to leave his females, and therefore no longer came to me when I passed: he only gave me at a distance many tokens of friendship, shouting and clapping his wings; but would not quit his companions, lest, perhaps, his rival should take possession. Things went on in this way till the breeding season, and he never gabbled to me but at a distance. When his females, however, began to sit, he left them, and redoubled his friendship to me. One day, having followed me as far as the ice-house at the top of the park, the spot where I must necessarily part with him in pursuing my way to a wood at half a league distance, I shut him in the park. He no sooner saw himself separated from me, than he vented strange cries. However, I went on my road; and had advanced about a third of the distance, when the noise of a heavy flight made me turn my head: I saw my Jacquot, only four paces from me. He followed me all the way, partly on foot, partly on wing; getting before me and stopping at the cross-paths to see which way I should take. Our journey lasted from ten

o'clock in the morning till eight in the evening;

windings of the wood, without seeming to be tired. After this he attended me every where, so as to become troublesome; for I was not able to go to any place without his tracing my steps, so that one day he even came to find me in the church. Another time, as he was passing by the rector's window, he heard me talking in the room; and, as he found the door open, he entered, climbed up stairs, and marching in, gave a loud exclamation of joy to the no small affright of the family.

" I am sorry, in relating such interesting traits of my good and faithful friend Jacquot, when I reflect that it was myself that first dissolved the pleasing connection; but it was necessary for me to separate him from me by force. Poor Jacquot fancied himself as free in the best apartments as in his own; and after several accidents of this kind, he was shut up and I saw him no more. His inquietude lasted above a year, and he died from vexation. He was become as dry as a bit of wood, as I am told, for I would not see him; and his death was concealed from me for more than two months after the event. Were I to recount all the friendly incidents between me and poor Jacquot, I should not for several days have done writing. He died in the year of our friendship, aged seven years and two months."

. A young goose is generally reckoned very good eating; yet the feathers of this bird still farther increase its value. Of goose feathers most of our beds in Europe are composed; in the countries bordering on the Levant, and in all Asia, the use of them is utterly unknown. They there use mattrasses, stuffed with wool, or camel's hair, or cotton; and the warmth of their climate may perhaps make them dispense with cushions of a softer kind. But how it happens that the ancients had not the use of feather-beds is surprising: Pliny tells us, indeed that they made bolsters of feathers to lay their heads on; and this serves as a proof that they turned feathers to no other uses.

The feathers of Somersetshire are most in esteem; those of Ireland are reckoned the worst. Hudson's Bay also furnishes very fine feathers, supposed to be of the goose kind.

THE BEAN GOOSE.

THIS is chiefly distinguished from the domestic goose by the resemblance of the nail of its bill to a horse-bean. The head and neck are of an ash brown, tinged with ferruginous; breast and belly dirty white; back, a plain ash-colour; feet and legs saffron, and claws black. They appear in the fens of Lincolnshire in autumn, from whence they migrate in May to the wild parts of Europe. While in this country they feed much on green wheat.

THE BERNACLE GOOSE.

THIS bird has a very short and black bill, crossed with a flesh-coloured mark on each side. Part of the head, the chin, throat, under parts, and upper tail-coverts, are white; and the rest of the head and neck, and the beginning of the back, are black; the thighs are mottled. Round the knee, the feathers are black; and the lower feathers of the back are the same, edged with white. The wing-coverts and scapulars are blue-grey; the ends black, fringed with white at the tip. The vent, tail, and legs, are black.

Bernacle geese are not uncommon on many of the northern and western coasts of this kingdom, in winter; but they are scarce in the south, and only seen in inclement seasons. They leave our island in February, and retire northward to breed.

Of all the marvellous productions which igno-

for the simple and truly wonderful operations of nature, the most absurd, and yet not the least celebrated, is the assertion of the growth of these birds, in a kind of shell, called lepas anatifera (goose-bearing shell), on certain trees on the coasts of Scotland and the Orkneys, or on the rotten timbers of old ships. For the entertain-ment of our readers, we shall give brief extracts from three, out of the numerous, writers who have credited these circumstances, and have all spoken positively upon the subject.

Maier, who has written a treatise expressly on this bird, says, " that it certainly originates from shells: and, what is still more wonderful, that he himself opened a hundred of the *goose-bearing* shells in the Orkneys, and found in all of them the rudiments of the bird completely formed."

Gerard, another writer on this point, and an Englishman, gives the following account of this wonderful transformation:—" What our eyes have seen, and our hands have touched, we shall declare. There is a small island in Lancashire, called the Pile of Foulders, wherein are found pieces of old and bruised ships, some whereof have been cast thither by shipwrecks; also the trunks and bodies, with the branches of old and rotten trees, cast up there likewise; whereon is found a certain spume or froth, that in time breedeth into certain shells, in shape like those of the muscle, but sharper pointed, and of a whitish colour, one end whereof is fastened unto the inside of the shell, even as the fish of oysters and muscles are; the other end is made fast unto the belly of a rude mass or lump, which in time

When it is perfectly formed, the shell gapeth

aforesaid lace or string; next cometh the legs of
the bird hanging out: and, as it groweth greater,
it openeth the shell by degrees, till at length it has
all come forth, and hangeth only by the bill. In
a short space after, it cometh to full maturity,
and falleth into the sea; where it gathereth feathers, and groweth to a fowl, bigger than a mallard and lesser than a goose, having black legs,
and bill or beak, and feathers black and white,
spotted in such manner as our magpie, called in
some places pie-annet, which the people of
Lancashire call by no other name than tree-
goose; which place aforesaid, and all those
places adjoining, do so much abound therewith,
that one of the best is bought for three-pence.
For the truth hereof, if any doubt, may it please
them to repair to me, and I shall satisfy them by
the testimonies of good witnesses."

Sir Robert Murray's account of the bernacle,

the following purport.—" In the Western Islands
of Scotland the West Ocean throws upon their
shores great quantities of very large weather-
beaten timber; the most ordinary trees are fir
and ash. Being in the island of East, I saw
lying upon the shore a cut of a large fir-tree, of
about two feet and a half in diameter, and nine
or ten feet long, which had lain so long out of

3

the water, that it was very dry; and most of the shells that had formerly covered it were worn or rubbed off. Only on the parts that lay next the ground, there still hung multitudes of little shells: they were of the colour and consistence of muscle-shells. This barnacle-shell is thin about the edges, and about half as thick as broad. Every one of the shells hath some cross seams or sutures, which, as I remember, divide it into five parts. These parts are fastened one to another, with such a film as muscle-shells are.

" These shells are hung at the trees by a neck, longer than the shell; of a kind of a filmy substance, round and hollow, and creased not unlike the windpipe of a chicken: spreading out broadest where it is fastened to the tree, from which it seems to draw and convey the matter which serves for the growth and vegetation of the shell and little bird within it.

" In every shell that I opened, I found *a perfect sea-fowl:* the little bill, like that of a goose; the eyes marked; the head, neck, breast, wing, tail, and feet, formed; the *feathers,* every where perfectly shaped and blackish-coloured; and the feet, like those of other water-fowl, to the best of my remembrance."

Few subjects seem to have been more circumstantially related, or to rest on better evidence, than the above; so natural to man is credulity, which passes all bounds where the prodigy of an event takes firm hold of the imagination, and

lays the understanding asleep. Such are the
wild chimeras that have been detailed concerning the origin of the bernacles; which ridiculous
fables once enjoyed great celebrity, and were
admitted by many authors. Such is the folly of
mankind to give credence to every wonderful
tale—such the dangerous contagion of error and
superstition!

The small bernacles frequent our coasts, as
well as those of Holland and Ireland, in winter;
they are of a brown colour, with the head, neck,
and breast, black, and a white collar. They are
easily tamed, and when fatted, are thought to be
delicate food. In some seasons they have been
known to resort to the coasts of France in such
numbers as to become a pest; and in the winter
of 1740, they destroyed all the corn near the sea
coasts, by tearing it up by the roots: a general
war was declared against them, and though thousands were knocked on the head, it availed but
little; nor were the inhabitants released from
this scourge till the north-wind, which brought
them, ceased to blow, when they took their
leave.

THE CANADA GOOSE.

THIS. bird is somewhat bigger than the tame
goose. The bill, the head, and the neck, are
black; and under the throat there is a broad

white band, like a crescent. The breast, the
upper part of the belly, the back, and wing-
coverts, are dusky brown; the lower parts of the
neck and belly, and upper tail-coverts, white.
The quills and tail are black, and the legs dark
lead-colour.

Canada geese inhabit the farther parts of
North America. Immense flocks appear annually in the spring in Hudson's Bay, and pass
more to the north to breed; and return southward in the autumn. The English at Hudson's
Bay depend greatly on geese, of this and other
kinds, for their support; and in favourable years
they often kill three or four thousand, which
they salt and barrel. Their arrival is impatiently
waited; it is the harbinger of the spring, and
that month is named by the Indians the Goose
Moon.

The English settlers send out their servants,
as well as the Indians, to shoot these birds on
their passage. It is in vain to pursue them; the
men therefore form a row of huts made of boughs,
at musket-shot distance from each other, and
placed in a line across the vast marshes of the
country. Each hovel, or, as it is called, stand,
is occupied by only a single person. These attend the flight of the birds; on the approach of
which they mimic their cackle so well, that the
geese will answer, wheel, and come nearer the
stand. The sportsman remains motionless, and
on his knees, with his gun cocked the whole

time; and never fires till he can perceive the
eyes of the goose. He fires as they are going
from him; then picks up another gun that lies
by him, and discharges that also. The geese
that he has killed, he sets up on sticks, as if alive,
to decoy others: he also makes artificial birds
for the same purpose. In a good day (for they
fly in very uncertain and unequal numbers) a
single Indian will kill two hundred. Notwithstanding each species of goose has a different
call, yet the Indians are admirable in their imi-
tation of every one,

THE SNOW GOOSE

IS about tne size of the common goose. The
upper mandible of the bill is scarlet, and the
lower one whitish. The general colour of the
plumage is white; except the first ten quills of
the wings, which are black with white shafts.
The young are of a blue colour, till they are a
year old. The legs are red.

These birds are very numerous about Hudson's Bay; where they are migratory, going farther northward to breed. They are also found
in the northern parts of the Old Continent.
About Jakut, and the other parts of Siberia, they
afford great subsistence to the natives, and the
feathers are an article of commerce. Each
family will kill thousands in a season, which,

after being plucked and gutted, are flung in heaps into holes dug for that purpose, and are covered only with earth. The mould freezes, and forms over them an arch, and whenever the family have occasion to open one of these magazines, they find their provisions perfectly sweet and good. In that frozen climate these birds have so little of the shyness of the other species, that they are taken in the most ridiculous manner imaginable. The inhabitants place near the banks of the rivers a great net, in a straight line; or else form a hovel of skins sewed together. This done, one of the company dresses himself in the skin of a white rein-deer, advances towards the flock of geese, and then turns back towards the net or hovel; and his companions go behind the flock, and by making a noise, drive them forward. The simple birds mistake the man in white for their leader, and follow him within reach of the net; which is suddenly pulled down, and captivates the whole. When he chuses to conduct them even into the hovel, they follow in the same manner; he creeps in at a hole left for that purpose, and out at another on the opposite side, which he closes up. The geese follow him through the first, and as soon as they are in, he passes round and secures every one of them. Mr. Hearne, however, in his " Journey to the Northern Ocean," says that the snow geese there are very different in their manners, being

5

the shyest and most watchful of all the species of geese, and never suffering a person to approach them within two or three gun-shots.

THE RACE-HORSE, OR, LOGGER-HEAD-GOOSE,

IS a large bird, weighing from twenty to thirty pounds; the bill is of an orange colour; the head, neck, and upper parts of the body of a deep ash-colour; the thighs inclining to blue; the quills and tail black, and on the bend of the wings, is a yellow knob, half an inch in length.

These birds are unable to fly, from the shortness of their wings, but make amazing progress on the water; their flesh is very rank and unsavory. They chiefly inhabit the Falkland Isles, Staten Land, &c. and are mostly seen in pairs.

THE RED-BREASTED GOOSE.

THIS is one of the most elegant of the race, though little known here, its residence being chiefly confined to the coasts of the Icy Sea; it generally weighs about three pounds, is quite free from any fishy taste, and therefore highly esteemed for the table.

THE RUDDY GOOSE

IS about the size of a mallard, and found in Siberia, from whence it migrates into India; its bill is black; the neck of an iron colour, encircled with a collar of black; the rest of the body an obscure or dusky red, except the tail, which is a greenish black. These birds frequently lay in hollow trees, and the male and female sit by turns; but all attempts to domesticate them have proved ineffectual. Their voice is not unlike the note of a clarinet. Their attachments are so very strong, that if the male be killed, the female will not quit the gunner till she has been two or three times shot at.

THE CYGNOIDES,

AS forming a middle line between the swan and the goose, have been not improperly stiled swan-geese. This species is the swan goose of Ray, from Guinea, and is also often called the Muscovy goose. They are frequent in Britain, and unite so readily with the common goose that their offspring will produce as certainly as if no such intermixture had taken place. They walk very erect, with the head much elevated; make an extraordinary harsh screaming noise, which they continue almost the whole day through, and without the least provocation or disturbance.

THE PUFFIN AUK

IS about the size of a teal, weighing near twelve ounces, and being about twelve inches in length. The eyes are ash-coloured, or grey; the upper part of the head and body are black; the lower parts white; it has a sort of black ring that encompasses the throat; the sides of the head are whitish, with a cast of yellow, or ash colour; the wings are made up of short feathers, and are very small: they fly swift while they keep near the surface of the water, on account of wetting their wings as they proceed. They have black tails, about two inches long; their legs and feet are of an orange colour, and their claws of a dark blue.

The bill is flat: but, very different from that of the duck, its edge is upwards. It is of a triangular figure, and ending in a sharp point; the upper chap bent a little downward, where it is joined to the head: and a certain callous substance encompassing its base, as in parrots. It is of two colours; ash-coloured near the base, and red towards the point. It has three furrows or grooves impressed in it; one in the livid part, two in the red. The eyes are fenced with a protuberant skin, of a livid colour; and they are grey or ash coloured.

The puffin, like the rest of the auk kind, has its legs thrown so far back, that it can hardly

5

move without tumbling. This makes it rise with
difficulty, and subject to many falls before it gets
upon the wing; but as it is a small bird, not much
bigger than a pigeon, when it once rises, it can
continue its flight with great celerity.

The puffin auks build no nest; but lay their
eggs either in the crevices of rocks, or in holes
under ground near the shore. They most generally chuse the latter situation. Relying on
its courage, and the strength of its bill, with
which it bites most terribly, it either makes or
finds a hole in the ground, where to lay or bring
forth its young. " All the winter," says Willoughby " these birds, like the rest, are absent;
visiting regions too remote for discovery. At the
latter end of March, or the beginning of April,
come over a troop of their spies or harbingers,
that stay two or three days, as it were to view
and search out for their former situations, and
see whether all be well. This done, they once
more depart; and about the beginning of May
return again with the whole army of their companions. But if the season happen to be stormy
and tempestuous, and the sea troubled, the unfor-
tunate voyagers undergo incredible hardships;
and they are found by hundreds, cast away upon
the shores, lean, and perished with famine. It
is most probable, therefore, that this voyage is
performed more on the water than in the air:
and as they cannot fish in stormy weather their

strength is exhausted before they can arrive at
their wished-for harbour."

The puffin, when it prepares for breeding,
(which always happens a few days after its arrival,) begins to scrape up a hole in the ground
not far from the shore, and when it has some
way penetrated the earth, it then throws itself
upon its back, and with its bill and claws thus
burrows inward, till it has dug a hole with several
windings and turnings, from eight to ten feet
deep. It particularly seeks to dig under a stone,
where it expects the greatest security. This is
chiefly the task of the males on which they are
so intent as to suffer themselves at that time to
be taken with the hand. Some, when there is
an opportunity, save themselves the trouble of
forming holes by dispossessing the rabbits of
theirs. In this fortified retreat the female lays
one white egg; which, though the bird be not
much bigger than a pigeon, is full the size of
that of a hen.

The males likewise perform the office of sitting, relieving their mates when they go to feed.
The young are hatched in the beginning of July.
When the young one is excluded, the parents'
industry and courage are incredible. Few birds
or beasts will venture to attack them in their retreats. When the great sea-raven, as Jacobson
informs us, comes to take away their young, the
puffins boldly oppose him. Their meeting af-

Battles between the puffin and raven.

fords a most singular combat. As soon as the
raven approaches, the puffin catches him under
the throat with its beak, and sticks its claw into
his breast, which makes the raven, with a loud
screaming, attempt to get away; but the little
bird still holds fast to the invader, nor lets him
go till they both come to the sea, where they
drop down together, and the raven is drowned:
yet the raven is but too often successful, and invading the puffin at the bottom of its hole, devours both the puffin and its family.

"But," Goldsmith observes with much propriety, "were a punishment to be inflicted for
immorality in irrational animals, the puffin is
justly a sufferer from invasion, as it is often itself
one of the most terrible invaders. Near the Isle
of Anglesey, in an islet called Priesholm, their
flocks may be compared, for multitude, to swarms
of bees. In another islet, called the Calf of Man,
a bird of this kind, but of a different species, is
seen in great abundance. In both places, numbers of rabbits are found to breed; but the puffin,
unwilling to be at the trouble of making a hole,
when there is one ready made, dispossesses the
rabbits, and it is not unlikely destroys their young.
It is in these unjustly acquired retreats that the
young puffins are found in great numbers, and
become a very valuable acquisition to the natives
of the place. The old ones (I am now speaking
of the Manks' puffin) early in the morning at
break of day, leave their nests and young, and

All this time they are diligently employed in fishing for their young; so that their retreats on land, which in the morning were loud and clamorous, are now still and quiet, with not a wing stirring till the approach of dusk, when their

ever fish, or other food, they have procured in the day, by night begins to suffer a kind of half digestion, and is reduced to an oily matter, which is ejected from the stomach of the old ones into the mouth of the young. By this they are nourished, and become fat to an amazing degree. When they are arrived to their full growth, they who are entrusted by the lord of the island, draw them from their holes; and, that they may more readily keep an account of the number they take, cut of one foot as a token. Their flesh is said to be excessively rank, as they feed upon fish, especially sprats and sea-weed; however, when they are pickled and preserved with spices, they are admired by those who are fond of high eating. We are told, that formerly their flesh was allowed by the church on Lenten days. They were at that time, also taken by ferrets, as we do rabbits. At present, they are either dug out, or drawn out, from their burrows, with an hooked stick. They bite extremely hard, and keep so fast hold of whatsoever they seize upon, as not to be easily disengaged. Their noise, when taken, is very disagreeable, being

like the efforts of a dumb person attempting
to speak.

" The constant depredation which these birds
annually suffer, does not in the least seem to intimidate them, or drive them away : on the contrary, the people say, the nest must be robbed,
or the old ones will breed there no longer. All
birds of this kind lay but one egg; yet if that
be taken away, they will lay another, and so on
to a third: which seems to imply, that robbing
their nests does not much intimidate them from
laying again. Those, howeevr, whose nests have
been thus destroyed, are often too late in bringing up their young; who, if they be not fledged
and prepared for migration when all the rest depart, are left at land to shift for themselves. In
August the whole tribe is seen to take leave of
their summer residence; nor are they observed
any more till the return of the ensuing spring.
It is probable that they sail away to more southern regions, as our mariners frequently see myriads of water-fowl upon their return, and steer-
ing usually to the north. Indeed the coldest
countries seem to be their most favored retreats;
and the number of water-fowl is much greater in
those colder climates, than in the warmer regions
near the line. The quantity of oil which abounds
in their bodies, serves as a defence against cold,
and preserves them in vigour against its severity;
but the same provision of oil is rather detrimental
in warm countries, as it turns rancid, and many

of them die of disorders which arise from its pu-
trefaction. In general, however, water-fowl can
be properly said to be of no climate; the element upon which they live, being their proper
residence. They necessarily spend a few months
of summer upon land, to bring up their young:
but the rest of their time is probably consumed
in their migrations, or near some unknown coasts,
where their provision of fish is found in the
greatest abundance."

Mr. Pennant has asserted, that their affection
for their young is so great, that when laid hold
of by the wings, they will give themselves the

they can reach, as if actuated by despair; and
when released, instead of flying away, they will
often hurry again into their burrows." However,
the Rev. Mr. Bingley informs us, " that when he
was in Wales, in the summer of 1801, he took
several of them out of the holes that had young
ones in them, for the purpose of ascertaining
this fact. They bit him with great violence,
but none of them seized on any parts of their
own body: a few, on being released, ran into

whence he had taken them: if it were more easy

selves into the air, they did so; but if not, they
ran down the slope of the hill in which their burrows were formed, and flew away. The noise
they make when with their young, is a singular

kind of humming, much resembling that produced by the large wheels used for spinning
worsted. On being seized, they emitted this
noise with greater violence; and from its being
interrupted by their struggling to escape, it
sounded not much unlike the efforts of a dumb
man to speak."

The young ones are entirely covered with a
long blackish down; and in shape are altogether
so different from the parent birds, that no one
could at first suppose them of the same species.
Their bill also is long, pointed, and black, with
scarcely any marks of furrows.

They feed on sprats or sea-weeds, which make
them excessively rank; yet the young are pickled
and preserved with spices, and by some people
are much admired.

The re-migration of the puffins takes place
about the middle of August; when not a single
one remains behind, except the unfledged young
of the latter hatches: these are left a prey to the
peregrine falcon; which watches the mouth of
the holes for their appearance, compelled, as
they must soon be, by hunger, to come out.

The Kamtschadales and Kuriles wear the bills
of the puffins fastened about their necks with
straps. The priests put them on with a proper
ceremony, and the persons are supposed to be
always attended by good fortune so long as they
retain them there.

From the observations made by the Rev.

Hugh Davies, of Aber, in Caernarvonshire, on the different forms of the bills, among the thousands of this species, which, in the year 1776, were wrecked on the Welsh coast near Criccieth, it appears certain that the puffins do not breed till their third year. He saw the beach, for miles, covered with dead birds; among which were puffins, razor-bills, guillemots, and kittiwakes; as well as tarrocks, gannets, wild geese, bernacles, brent geese, scoters, and tufted ducks. This unusual accident he conjectured to be owing to a severe storm of frost that had overtaken both the migrants and re-migrants. From the puffins he here found, he remarked the different forms of their bills in their several periods of life. Those that he supposes to have been of the first year, were small, weak, destitute of any furrow, and of a dusky colour; those of the second year were considerably stronger and larger, lighter coloured, and with a faint rudiment of a furrow at the base; those of the more advanced years had vivid colours, and were of great strength.

THE NORTHERN DIVER,

WHICH is the principal of the auk tribe, is is black; and is four inches and a half long. The head and neck are of a deep velvet black.

Description.

Under the chin is a patch of white, marked with
several parallel lines of black; and on each side
of the neck, and on the breast, is also a large
portion of white marked in a similar manner.
The upper parts are black, marked with white
spots; and the under parts are white. The wings
are short: and the quills, tail, and legs are black.
The female is less than the male. It inhabits
chiefly the northern seas; and is common on
some of the coasts of Scotland.

Every part and proportion of this bird is most
admirably adapted to its mode of life. The head
is sharp; and smaller than the part of the neck
adjoining, in order that it may pierce the water:
the wings are placed forward, and out of the
centre of gravity, for a purpose which will be
noticed hereafter: the thighs quite at the podex,
in order to facilitate diving: and the legs are
flat, and as sharp backwards almost as the edge
of a knife, that, in striking, they may easily cut
the water; while the feet are broad for swimming, yet so folded up, when advanced forward
to take a fresh stroke, as to be full as narrow as
the shank. The two exterior toes of the feet are
longest: and the nails are flat and broad, resembling the human; which give strength, and increase the power of swimming. The foot, when
expanded, is not at right angles with the leg;
but the exterior part, inclining towards the head,
forms an acute angle with the body, the inten-
tion being, not to give motion in the line of the

legs themselves, but, by the combined impulse of both, in an intermediate line, the line of the body.

Most people who have exercised any degree of observation, know that the swimming of birds is nothing more than a walking in the water, where one foot succeeds the other as on the land; " but no one, as far as I am aware," says the Rev. Mr. White, " has remarked that diving-fowls, while under water, impel and row themselves forward by a motion of their wings, as well as by the impulse of their feet: yet such is really the case, as any one may easily be convinced who will observe ducks when hunted by dogs in a clear pond. Nor do I know that any one has given a reason why the wings of diving-fowls are placed so forward: doubtless, not for the purpose of promoting their speed in flying, since that position certainly impedes it; but probably for the increase of their motion under water, by the use of four oars instead of two; and were the wings and feet nearer together, as in land-birds, they would when in action, rather hinder than assist one another."

THE SPECKLED DIVER.

THIS is not quite so large as the preceding; it has a straight sharp bill, of a sort of livid or black colour, with feathers growing down as low as the nostrils, that part of the neck next to the head is covered with feathers set so exceedingly thick, that it looks as large as the head itself, the lower parts of the body are white; the upper parts of a dusky sort of dark grey, speckled over with white spots, which are larger upon the wings than on the rest of the body; the fore toes are very long, especially the outermost, the back toes are but little, and short, the legs of a brown colour, and not very long, and are placed so back, that the bird seems scarce able to walk without erecting itself perpendicularly on its tail which is very short. Some of these birds hav a sort of ring, about their necks, with blacker heads, and sprinkled with little white specks, and lines; others are more grey, or ash-coloured, and varied with white specks, but no lines, which may perhaps be the distinction between the cocks and the hens.

THE SEA DIVER

WEIGHS about three pounds; the bill is upward of two inches and a half long; the whole body is covered with fine soft thick feathers, the head and neck of a brown colour, but the back darker, each side or the body more dusky, the breast and belly inclining pretty much to a silver colour. It has not any tail at all; the outermost quill feathers of the wings, are blackish, the lesser rows underneath are white.

The bill appears compressed sideways, and is narrow, of a reddish colour; the tongue a little cloven, the eyes dark with a sort of red mixture. The claws appear broad, resembling in some degree the nails of a man's hand, on one side quite black, the other of a pale blue, or rather of an ash colour, the outermost toe longer than the rest, both the legs and toes are broad and flat. It feeds on small fishes, sea weeds, &c.

THE CRESTED DIVER

IS about the size of a duck; the bill, that part especially towards the head is of a reddish colour, and in length is something more than two inches; on the top of the head and neck, is a beautiful crest of feathers, those on the neck appearing like a collar or ruff, and seem a good deal bigger than they really are; those on the top of the head are black, those on the sides of the neck, are of a reddish or cinerous colour; the back and wings are of a darkish brown, pretty much inclining to black, except some of the exterior edges of the wing feathers, which are white. The breast and belly are of a light ash-colour; it has no tail; the legs and toes broad, and flat, much like those before described. It has an unpleasant cry, and will occasionally, when angered or pleased, raise or fall the feathers of his crest.

THE GREAT AUK.

THIS bird is the size of a goose; its bill is black, about four inches and a quarter in length, and covered at the base with short velvet-like feathers. The upper parts of the plumage are black, and the lower parts white, with a spot of white between the bill and the eyes, and an oblong stripe of the same on the wings, which are too short for flight. The bird is also a very bad walker, but swims and dives well. It is, however, observed by seamen, that it is never seen out of soundings, so that its appearance serves as an infallible direction to land. It feeds on the lump-fish, and others of the same size; and is frequent on the coast of Norway, Greenland, Newfoundland, &c. It lays its eggs close to the sea mark.

THE RAZOR-BILL,

WHICH is not so large as the common tame duck, has a large bill of a deep black colour and near two inches long; with a deep incision or furrow in the upper mandible, which runs a little beyond the nostrils, and is in some degree covered with a sort of nappy thick down-like velvet; the upper part being crooked at the end and hanging over the under, with transverse channelled lines running across each, and a narrow white line passing from each eye to the corner of the upper mandible. The inner part of the mouth is of a fine yellow, and the eyes of hazel colour. The head and upper part of the body are black: the under part of the chin more purple, the breast, belly, and tips of the covert feathers of the wings white. The tail is black, and about three inches long; the legs, feet, and toes pretty much of the same colour. They breed on the edges of steep craggy rocks, by the sea-shore, laying large white eggs spotted with black.

THE GUILLEMOT

IS about the size of a common duck, the upper parts of the body are of a dark brown colour, inclining to a black, except the tips of some of the wing feathers which are white; all the under parts of the body are also white. The tail is about two inches long.

. These are simple birds, and easily taken. They generally join companies with other birds, and breed on the inaccessible rocks, and steep cliffs in the Isle of Man, and likewise in Cornwall; on

Prestholm Island, near Beaumaris in the Isle of Anglesey; also on the Fern Island, near Northumberland, and in the cliffs about Scarborough, in Yorkshire, and several other places in England. They lay exceeding large eggs, being full three inches long, blunt at one end, and sharp at the other, of a sort of bluish colour, spotted generally with some black spots or strokes.

THE LESSER GUILLEMOT

WEIGHS about sixteen ounces. The upper parts of its plumage are darker than those of the former species. The black guillemot is entirely black, except a large mark of white on the wings. In winter, however, this bird is said to change to white; and there is a variety in Scotland not uncommon, which is spotted, and which Mr. Edwards has described under the name of the spotted Greenland dove. The marbled guillemot, which is found at Kamtschatka, &c. received its name from its plumage, which is dusky, elegantly marbled with white.

GOOSEANDER AND DUN DIVER.

THIS is the largest of the auk kind, weighing about four pounds. The bill is red; the head is very full of feathers on the top and back part. The plumage is various and beautiful. The head and upper parts are of a fine glossy black, the rump and tail ash-colour, and the under parts of the neck and body a fine pale yellow. Its manners and appetites entirely resemble those of the diver. It feeds upon fish, for which it dives; and it is said to build its nest upon trees, like the cormorant.

The dun 'diver is less than the gooseander. The upper part of the head is reddish brown; the back and wings ash colour, and the lower part of the body white. It is found in the same places, and has the same manners with the gooseander.

HOODED, AND RED-BREASTED MERGANSER

IS a native of North America. It is about the size of a wigeon. The head and neck are dark brown, the former surrounded with a large round crest, the middle of which is white. The back and quills are black, the tail dusky; and the breast and belly white, undulated with black. The female is fainter in the colour of her plumage, and has a smaller crest.

The red-breasted merganser weighs only two pounds. The head and neck are black, glossed with green, the rest of the neck and the belly white; the upper part of the back is glossy black; the lower parts and the rump are striated with brown and pale grey; on the wings there are white bars tipped with black, and the breast is reddish, mixed with black and white. The plumage of the female is less splendid; and they differ in another respect, viz. that the male has a very full and large crest; the female only the rudiment of one.

THE SMEW

MEASURES from the end of the bill to the end of the tail near eighteen inches, and from the extremity of each wing when extended, upwards of two feet, and weighs about a pound and a half. It has a fine crest upon the head, which falls down towards the back part of it, under which, on each side of the head, is a black spot; the rest of the head and the neck are white, as is the under parts of the body; the back and the wings are of an agreeable mixture of black and white. The tail is about three inches long, of a sort of dusky ash-colour, the feathers on each side shortening gradually. The bill is of a lead colour, at the extremity of which is a dirty coloured spot of white; it is something less than the generality of the duck kind, a little hooked with large open nostrils; and darkish coloured eyes; the legs are pretty much of the same colour of the bill.

The female of this bird has no crest; the sides of the head are red, the throat white, the wings of a dusky ash-colour; in other respects it agrees with the male. They feed on fish, but are very rarely seen in England, except in very hard seasons, and then not more than three or four of them together.

THE CHINESE DIVER.

THE bill of this bird is said to be dusky. The upper parts of the plumage are greenish brown; and the fore-part of the neck the same, but paler. The chin, and under parts, are yellowish white, marked with dusky spots. The legs are ash-coloured. Its size has never been ascertained.

This is supposed to be one of the birds used by the Chinese for catching fish. In that employment it has a ring fastened round the middle of the neck, to prevent its swallowing; it has also a long slender string fastened to it; thus accoutred, it is taken by its master into the fishing-boat, from the edge of which it is taught to plunge after the fish as they pass by; and as the ring prevents these from passing down into the throat, they are taken from the mouth of the bird as fast as it catches them. In this manner it frequently happens that a great many are procured in the course of a few hours. When the keeper has taken a sufficient quantity of fish for himself, the ring is taken off, and the poor

labourer is suffered to satisfy its own hunger.
The bird most commonly used for this purpose
by the Chinese fishermen is a species of pelican,
called the fishing cormorant; when treating of
which we shall give a further account of this singular mode of fishing.

THE PENGUIN.

THE penguins seem to hold the same place
in the southern parts of the world, that the auks
do in the northern; being only found in the temperate and frigid zones of the southern hemi-
sphere. They resemble them in almost all their
habits; walking erect, and being very stupid:
they also resemble them in their colour, and
their mode of feeding, and of making their nests.
They hatch their young in an erect position;
and cackle like geese, but in a hoarser tone.

The bill of the penguin is strong, straight,
furrowed on the sides, and bent towards the
point. The nostrils are linear, and placed in the
furrows. The tongue is covered with strong
spines, pointing backwards. The body is clothed
with thick short feathers; which have broad
shafts, and are placed as compactly as scales.
The legs are short and thick, placed backwards,
near the tail. The toes are four, all placed for-
wards; the interior ones are loose, and the rest
webbed. The tail is very stiff, consisting of

broad shafts scarcely webbed. The wings are small, not unlike fins, covered with no longer feathers than the rest of the body. Their wings serve them rather as paddles to help them forward, when they attempt to move swiftly; and in a manner walk along the surface of the water. Even the smaller kinds seldom fly by choice; they flutter their wings with the swiftest effort without making way; and though they have but a small weight of body to sustain, yet they seldom venture to quit the water where they are provided with food and protection.

Though the wings of the penguin tribe be unfitted for flight, their legs are still more aukwardly adapted for walking. The whole tribe have all above the knee hid within the belly; and nothing appears but two short legs, or feet, as some would call them, that seem stuck under the rump, and upon which the animal is very aukwardly supported. They seem, when sitting, or attempting to walk, like a dog that has been taught to sit up, or to walk on his hind legs. Their short legs drive the body in progression from side to side; and were they not assisted by their wings, they could scarcely move faster than a tortoise.

This aukward position of the legs, which so unqualifies them for living upon land, adapts them admirably for a residence in water. In that, the legs placed behind the moving body, pushes it forward with greater velocity; and

Admirably adapted for swimming and diving
___

these birds, like Indian canoes, are the swiftest
in the water, by having their paddles in the rear.
Our sailors, for this reason, give these birds a
very homely, but at the same time, expressive
name.

Nor are they less qualified for diving than
swimming. By ever so little inclining their bodies forward, they lose their centre of gravity;
and every stroke from their feet only tends to
sink them the faster. In this manner they can
either dive at once to the bottom, or swim between two waters; where they continue fishing
for some minutes, and then ascending, catch an
instantaneous breath, to descend again to renew
their operations. Hence it is that these birds
which are so defenceless, and so easily taken by
land, are impregnable by water. If they perceive themselves pursued in the least, they instantly sink, and show nothing more than their
bills, till the enemy is withdrawn. Their very
internal conformation assists their powers in
keeping long under water. Their lungs are fitted
with numerous vacuities, by which they can take
in very large inspiration; and this probably
serves them for a length of time.

As they never visit land, except when they
come to breed, the feathers take a colour from
their situation. That part of them which has
been continually bathed in the water, is white;
while their backs and wings are of different
colours, according to the different species. They

are also covered more warmly all over the body with feathers, than any other bird whatever; so that the sea seems entirely their element; and but for the necessary duties of propagating the species, it is probable we should scarcely have the smallest opportunity of seeing them, and should be utterly unacquainted with their history.

The penguins generally lay their eggs in holes in the ground. In some countries, however, they nestle in a very different manner, and which most of our naturalists ascribe to the frequent disturbances it has received from man or quadrupeds, in its former recesses. In some places, instead of contenting itself with a superficial depression in the ground, the penguin is found to burrow two or three yards deep; in others it is seen to forsake the level, and to clamber up the ledge of a rock, where it lays its egg, and hatches it in that bleak, exposed situation; and which precautions most probably have been adopted in consequence of dear-bought experience. In those countries where the bird fears for her own safety, or that of her young, she may providentially provide against danger, by digging, or even by climbing; for both which she is but ill adapted by nature. In those places however, where the penguin has had but few visits from man, her nest is made, with the most confident security, in the middle of some large plain where they are seen by thousands. In that unguarded situation, neither expecting nor fearing a power-

5

ful enemy, they continue to sit brooding; and even when man comes among them, have at least no apprehension of their danger. " But it is not considered," observes a judicious writer, " that these birds have never been taught to know the dangers of an human enemy; it is against the fox or the vulture that they have learned to defend themselves; but they have no idea of injury from a being so very unlike their natural opposers. The penguins, therefore, when our seamen first came amongst them, tamely suffered themselves to be knocked on the head, without even attempting an escape. They have stood to be shot at in flocks, without offering to move, in silent wonder, till every one of their number has been destroyed. Their attachment to their nests was still more powerful; for the females tamely suffered the men to approach and take their eggs, without any resistance. But the experience of a few of these unfriendly visits, has long since taught them to be more upon their guard in chusing their situations; or to leave those retreats where they were so little able to oppose their invaders."

The penguin lays but one egg; and, in frequented shores is found to burrow like a rabbit: sometimes three or four take possession of one hole, and hatch their young together. In the holes of the rocks, where nature has made them a retreat, several of this tribe, as Linnæus assures us, are seen together. There the females lay their single egg in a common nest, and sit upon this their general possession by turns; while one is placed as a sentinel, to give warning of approaching danger. The egg of the penguin, as well as of all this tribe, is very large for the size of the bird, being generally found bigger than that of a goose. But as there are many varieties of the penguin, and as they differ in size, from that of a muscovy duck to a swan, the eggs differ in the same proportion.

As far as it is at present known, the penguins consist of about nine species, and they are commonly estimated to hold the same place in the southern parts of the world as the auks do in the north, neither of them having been observed within the tropics.

The one commonly denominated the Patagonian penguin, is by much the largest, some of them weigh at least forty pounds, and are four feet three or four inches in length. The bill measures four inches and a half, but is slender. The head, throat, and hind part of the neck is brown, the back of a deepish ash-colour, and all the under parts white. The best known penguin is not bigger than a common goose, the upper parts of whose plumage is black, and the under white.

`THE MEGALLANIC PENGUIN.

THIS is the most singular and remarkable of all the penguin tribe. In size it approaches that of a tame goose. It never flies, as its wings are very short, and covered with stiff hard feathers, which are always seen expanded, and hanging useless down by the bird's sides. The upper part of the head, back, and rump, are covered with stiff, black feathers; while the belly and breast, as is common with all of this kind, are of a snowy whiteness, except a line of black that is seen to cross the crop. The bill, which from the base to about half way is covered with wrinkles, is black, but marked crosswise with a stripe of yellow. They walk erect with their heads on high, their fin-like wings hanging down like arms; so that to see them at a distance, they look like so many children with white aprons. From hence they are said to unite in themselves the qualities of men, fowls, and fish. Like men, they are upright; like fowls they are feathered;-and like fish they have fin-like instruments, that beat the water before, and serve for all the purposes of swimming rather than flying.

They feed upon fish: and seldom come ashore, except in the breeding season. As the seas in that part of the world abound with a variety, they seldom want food; and their extreme fatness seems a proof of the plenty in which they

live. They dive with great rapidity, and are voracious to a great degree. One of them described by Clusius, though but very young,
would swallow an entire herring at a mouthful,
and often three successively before it was ap-
peased. In consequence of this gluttonous appetite, their flesh is rank and fishy; though the
English sailors say, " that it is pretty good eating." In some, the flesh is so tough, and the
feathers so thick, that they stand a blow of a scy-
mitar without injury.

They are birds of society; and especially when
they come on shore, they are seen drawn up in
rank and file, upon the ledge of a rock, standing
together with the albatross, as if in consultation.
This is previous to their laying, which generally
begins in that part of the world in the month
of November. Their preparations for laying are
attended with no great trouble, as a small depression in the earth, without any other nest, serves
for this purpose. The warmth of their feathers
and of their bodies is such, that the progress of
incubation is carried on very rapidly.

THE CRESTED PENGUIN,

WHICH is the most beautiful of the penguin
tribe, is nearly two feet in length. The bill is
red, and three inches long; the upper mandible
curved at the end, and the lower obtuse. The

2

head, neck, back, and sides, are black. Over
each eye there is a stripe of pale yellow feathers,
which lengthens behind into a crest about four
inches long; this is decumbent, but can be
erected at pleasure; the feathers of the head
above this are longer than the rest, and stand upward. The wings are black on the outside: but
the edges and the inside are white. The legs are
orange-coloured, and the claws dusky. The female is destitute of the crest. These birds have
also the names of hopping penguins, and jump-
ing jack, from their action of leaping quite out
of the water, sometimes three or four feet, on
meeting with any obstacle in their course; and,
indeed, they frequently do this without any other
apparent cause than the desire of advancing by
that means. They are inhabitants of several of
the South-Sea islands.

This species seems to have a greater air of
liveliness in its countenance than almost any of
the others: yet it is a very stupid bird; and
so regardless of its own safety, as even to suffer
any person to lay hold of it. When provoked,
it erects its crest in a very beautiful manner; and
when attacked by our voyagers, we are told it
ran at them in flocks, pecked their legs, and
spoiled their clothes. Mr. Forster, in his account of one of the South-Sea islands, says,
" When the whole herd was beset, they all became very bold at once; and ran violently at us,
biting our legs, or any part of our clothes."

They are very tenacious of life. Mr. Forster left a great number of them apparently lifeless from the blows they had received, while he went in pursuit of others; but they all afterwards got up and marched off with the utmost gravity.

Their sleep is extremely sound: for Dr. Sparrman accidentally stumbling over one of them, kicked it several yards without disturbing its rest; nor was it till after being repeatedly shaken that the bird awoke.

The crested penguins form their nests among those of the birds of the pelican tribe, and live in tolerable harmony with them. The female generally lays only a single egg. Their nests are holes in the earth; which they easily form with their bills, throwing back the dirt with their feet. They are often found in great numbers on the shores where they have been bred.

Penrose, in his " Account of an Expedition to the Falkland Islands in 1772," mentions a species of penguin that resorts to certain places of these islands in incredible numbers, and lays its eggs. These places, he says, had become by its long residence entirely freed from grass; and he has given to them the name of towns. The nests were composed of mud; raised into hillocks, about a foot high, and placed close to each other. " Here," he adds, " during the breeding season, we were presented with a sight that conveyed a most dreary, and, I may say, awful idea of the desertion of the islands by the human species :—

general stillness prevailed in these towns; and whenever we took our walks among them, in order to provide ourselves with eggs, we were regarded, indeed, with side-long glances, but we carried no terror with us.

" The eggs are rather larger than those of a goose, and are laid in pairs. When we took them once, and sometimes twice in a season, they were as often replaced by the birds; but prudence would not permit us to plunder too far, lest a future supply in the next year's brood might be prevented."

There is also a species at New Zealand not larger than a teal, and in almost all parts of the South Seas they are found in abundance, of all kinds and sizes.

CHAP. IX.

"Others there are—voracious—bold—
With pouches curious to behold—
Than which, no greater foes can be
To all the natives of the sea."

ANON.

THE PELICAN.

THESE birds are gregarious, and in general notorious for their extreme voracity. The white pelican (which is the most remarkable of

tending to eight or nine inches down the neck; this is bare of feathers, and is capable of containing many quarts of water. The tongue is so small as to be scarcely distinguishable. The sides of the head are naked; and on the back of the head is a kind of crest. The whole plumage is whitish, suffused with a pale blush-colour; ex-

cept some parts of the wings, which are black.
The legs are lead-coloured, and the claws grey.

The bag in the lower mandible of the bill is
one of the most remarkable members that is found
in the structure of any animal. Though it
wrinkle up nearly into the hollow of the chap,
and the sides to which it is attached are not (in
a quiescent state) above an inch asunder, it may
be distended amazingly; and when the bird has
fished with success, its size is almost incredible.
It would contain a man's head with the greatest
ease; and it has even been said that a man's leg,
with a boot on, has been hidden in one of these
pouches. In fishing, the pelican fills this bag:
and does not immediately swallow his prey: but
when this is full, he returns to the shore to devour
at leisure the fruits of his industry. He is not
long in digesting his food; for be has generally
to fish more than once in the course of a day.
When the bill of this bird is opened to its widest
extent, a person may run his head into the bird's
mouth, and conceal it in this monstrous pouch,
thus adapted for very singular purposes. Yet
this is nothing to what Ruysch assures us, who
avers, that a man has been seen to hide his whole
leg, boot and all, in the monstrous jaw of one of
these animals. At first appearance this would
seem impossible, as the sides of the under chap,
from which the bag depends, are not above an
inch asunder when the bird's bill is first opened;
but then they are capable of great separation;

and it must necessarily be so as the bird preys upon the largest fish, and hides them by dozens in its pouch. Father Tertre affirms, that it will hide as many fish as will serve sixty hungry men for a meal.

These birds, who are very lazy and indolent when they have glutted themselves with fish, retire when the toils of the day are over, a little way on the shore to take their rest for the night. They remain almost motionless, if not fast asleep, till hunger calls them to break off their repose; thus spending nearly the whole of their life in eating and sleeping. When thus incited to exertion, they fly from the spot, and, raising themselves thirty or forty feet above the surface of the sea, turn their head with one eye downwards, and continue to fly in that posture till they see a fish sufficiently near the surface: they then dart down with astonishing swiftness, seize it with unerring certainty, and store it up in their pouch. Having done this, they rise again; and continue the same actions till they have procured a competent stock.

The ancients, who were particularly fond of the marvellous, almost unanimously agreed in giving this bird admirable qualities and parental affections; struck, perhaps, with this extraordinary figure, they were willing to supply it with as extraordinary appetites; and having found it with a large reservoir, they were pleased with turning it to the most tender and parental uses. But we have the authority of the great Buffon for asserting that the pelican is a very sluggish voracious bird, and very ill fitted to take those flights, or make those cautious provisions for a distant time, which the ancients were so partial of attributing to them. Father Labat, who seems to have studied their manners with great exactness, has given a minute history of this bird, as found in America.

"The pelican," says he, "has strong wings, furnished with thick plumage of an ash-colour, as are the rest of the feathers over the whole body. Its eyes are very small, when compared to the size of its head; there is a sadness in its countenance, and its whole air is melancholy. It is as dull and reluctant in its motions, as the flamingo is sprightly and active. It is slow of flight; and when it rises to fly, performs it with difficulty and labour. Nothing, as it would seem, but the spur of necessity, could make these birds change their situation, or induce them to ascend into the air: but they must either starve or fly.

"They are idle and inactive to the last degree, so that nothing can exceed their indolence but their gluttony; it is only from stimulations of hunger that they are excited to labour; for otherwise they would continue always in fixed repose. When they have raised themselves about thirty or forty feet above the surface of the sea, they turn their head with one eye downwards, and continue to fly in that posture.. As soon as they perceive a fish sufficiently near the surface, they dart down upon it with the swiftness of an arrow, seize it with unerring certainty, and store it up in their pouch. They then rise again, though not without great labour, and continue hovering and fishing, with their head on one side as before.

" This work they continue with great effort and industry till their bag is full, and then fly to land to devour and digest at leisure the fruits of their industy. This, however, it would appear they are, not long in performing; for towards night they have another hungry call; and they again reluctantly go to labour. At night, when their fishing is over, and the toil of the day crowned with success, these lazy birds retire. a little way from the shore; and though with the webbed feet and clumsy figure of a goose, they will be contented to perch no where but upon trees among the light and airy tenants of the forest. There they take their repose for the night; and often spend a great part of the day, except such times as they are fishing, in dismal solemnity, and as it would seem half-asleep. Their attitude is with the head resting upon their great bag, and that resting upon their breast. There they remain without motion, or once changing their situation, till the calls of hunger break their repose, and till they find it indispensibly neces-

sary to fill their magazine for a fresh meal. Thus
their life is spent between sleeping and eating;
and being as foul as they are voracious, they are
every moment voiding excrements in heaps as
large as one's fist."

"The same indolent habits," says another author, "seem to attend them even in preparing for
incubation, and defending their young when excluded. The female makes no preparation for
her nest, nor seems to chuse any place in preference to lay in; but drops her eggs on the bare
ground to the number of five or six, and there
continues to hatch them. Attached to the place,
without any desire of defending her eggs or her
young, she tamely sits and suffers them to be
taken from under her. Now and then she just
ventures to peck, or to cry out when a person
offers to beat her off."

The pelican, however, is by no means destitute
of natural affection, either towards its young, or
towards others of its own species, as will appear
from the annexed engraving. Clavigero, in his
History of Mexico, says that some of the Americans, in order to procure a supply of fish without any trouble, cruelly break the wing of a live
pelican, and after tying the bird to a tree, conceal themselves near the place. The screams of
the miserable bird attract other pelicans to the
place, which, he assures us, eject a portion of the
provisions from their pouches for their impri-
soned companion: as soon as the men observe

this, they rush to the spot, and, after leaving a small quantity for the bird, carry off the remainder.

The female feeds her young with fish macerated for some time in her bag; and when they cry, flies off for a new supply. Labat informs us, that he took two pelicans when very young, and tied them by the leg to a post stuck into the ground; and he had the pleasure of seeing the old one come for several days to feed them, remaining with them the greatest part of the day, and spending the night on the branch of a tree that hung over them. They all three thus became so familiar as to suffer themselves to be handled; and the young ones always took the fish that he offered to them, storing it first in their bag, and then swallowing it at leisure.

The pelican has often been rendered entirely

one among the Americans so well trained that it would, on command, go off in the morning, and return before night with its pouch distended with prey, part of which it was made to disgorge, and the rest it was permitted to retain for its trouble.

According to the account of Faber, a pelican was kept in the court of the Duke of Bavaria above forty years. He says that it seemed very

when any one sang or played on an instrument, it would stand perfectly still, turn its ear to the

place, and, with its head stretched out, seem to pay the utmost attention. Gessner informs us, that the emperor Maximilian had a tame pelican that lived about eighty years, and always attended his soldiers when on their marches. It was one of the largest of the kind, and had a daily allowance by the emperor's orders. M. de Saint Pierre mentions his having seen at the Cape Town a large pelican, playing close to the custom-house with a great dog; whose head she often took, in her frolic, into her enormous beak.

When a number of pelicans and cormorants are together, they are said to have a very singular method of taking fish. They spread into a large circle, at some distance from land; and the pelicans flap with their extensive wings above, on the surface, while the cormorants dive beneath: hence the fish contained within the circle are driven before them towards the land; and, as the circle lessens by the birds coming closer together, the fish, at last, are brought into a small compass, when their pursuers find no difficulty in filling their bellies. In this exercise they are often attended by various species of gulls, who likewise obtain a share of the spoil.

THE ALBATROSS.

THIS is one of the largest and most formidable birds of Africa and America, but as yet few

opportunities have occurred whereby the particulars of its natural history could be obtained, and
therefore that little which has been mentioned
must be subject to some doubt. In speaking of its
figure, Edwards appears to be the most correct;
he says, " the body is rather larger than that of
the pelican; and its wings, when extended, ten
feet from tip to tip. The bill, which is six inches
long, is yellowish, and terminates in a crooked
point. The top of the head is of a bright brown;
the back is of a dirty deep spotted brown; and
the belly and under the wings is white: the toes,
which are webbed, are of a flesh colour."

Such are the principal traits which this author
points out in the albatross's figure; but of any
peculiarities in its manners and disposition, which
might lead us to some knowledge of its history,
the greatest part of our naturalists have been entirely silent. A bird has however been described
by Wicquefort, under the title of the Alcatraz,
which, from its size, colours, and choice of its
prey, leaves no room to doubt of its being the same
we have under consideration. He describes it
as a kind of great gull, as large in the body as a
goose, of a brown colour, with a long bill, and
living upon fish, of which they kill great numbers.

" This bird is an inhabitant of the tropical climates, and also beyond them as far as the Straits
of Magellan in the South Seas. It is one of the
most fierce and formidable of the aquatic tribe,
not only living upon fish, but also such small

water-fowl as it can take by surprise. It may be considered as the principal of the gull kind; like the whole of them it seeks its prey upon the wing; and chiefly pursues the flying fish, that are forced from the sea by the dolphins. The ocean in that part of the world present a very different appearance from the seas with which we are sur-rounded. In our seas we see nothing but a dreary expanse ruffled by winds, and seemingly forsaken by every class of animated nature. But the tropical seas, and the distant southern latitudes beyond them, are all alive with birds and fishes, pursuing and pursued. Every various species of the gull kind are there seen hovering on the wing, at an immense distance from the shore. A picture of which has been thus most ably drawn:—" the flying fish are every moment rising to escape from their pursuers of the deep; only to encounter equal dangers in the air. Just as they rise the dolphin is seen to dart after them, but generally in vain; the gull has more frequent success, and often takes them at their rise; while the albatross pursues the gull, and obliges it to relinguish its prey: so that the whole horizon presents but one living picture of rapacity and invasion."

. These facts have been too clearly ascertained to leave the smallest kind of doubt; but how far credence may be given to Wicquefort, in what he adds concerning the albatross, the reader is left to determine; " these birds," says that wri-ter, " except when they breed, live entirely remote from land, so they are often seen, as it should seem, sleeping in the air. At night, when they are pressed by slumber, they rise into, the clouds as high as they can; there putting their head under one wing, they beat the air with the other, and seem to take their ease. After a time, however, the weight of their bodies, only thus half supported, brings them down; and they are seen descending with a pretty rapid· motion, to the surface of the sea. Upon this they again put forth their efforts to rise; and thus alternately ascend and descend at their ease. But it sometimes happens, that, in their slumbering flights, they are off their guard, and fall upon deck, when they are taken."

Be this account true or false, it is certain, that few birds float upon the air with more ease than the albatross, or support themselves a longer time in that element. They seem never to feel the excesses of fatigue; but night and day upon the wing are always prowling, yet always emaciated and hungry.

Though this bird is of the most voracious disposition, and thus tyrannical in its nature, yet he is also a proof that there are some associates which even tyrants themselves form, to which they are induced either by caprice or necessity. The albatross seems to have a peculiar affection fot the penguin, and a pleasure in its society. They are always seen to chuse the same places of breeding—some distant uninhabited island, where the ground slants to the sea, as the penguin is not formed either for flying or climbing. In such places their nests are seen together, as if they stood in need of mutual assistance and protection. Captain Hunt, who for some time commanded at our settlement upon Falkland Islands, has declared, that he was often amazed at the union preserved between these birds, and the regularity with which they built together. In that bleak and desolate spot, where the birds had long continued undisturbed possessors, and no way dreaded the encroachments of men, they seemed to make their abode as comfortable as they expected it to be lasting. They were seen to build with an amazing degree of uniformity; their nests, by thousands, covering fields, and resembling a regular plantation. In the middle, on high, the albatross raised its nest, on heath sticks and long grass, about two feet above the surface: round this the penguins made their lower settlements, rather in holes in the ground; and most usually eight penguins to one albatross. Nothing is a stronger proof of M. Buffon's fine observation, that the presence of man not only destroys the society of meaner animals, but their instincts also; for we have it as a positive fact, that these nests are now totally destroyed; the society is broke up, and the albatross and penguin have gone to breed upon more desert shores, where they conceive themselves safe from his intrusion, and where they can securely preserve that peace and safety, which he is constantly sure to interrupt.

MAN-OF-WAR BIRD, OR FRIGATE,

SO called, on account of the swiftness of its
flight, and its large and spreading wings, is found
in most seas, but chiefly in those within the tropics: it is, however, often seen about the Cape
of Good .Hope; and towards the end of July,
these birds collect in great numbers in Kamtschatka, and the seas which separate that part of
Asia from America. In size, the man-of-war is
sometimes as large as a swan; its breast is very
fleshy, and its general colour is white, the upper
parts marked with black lines. The quill-feathers
are black; and the tail is rounded, and of a lead
colour. The bill is of a pale yellow, and the legs
are flesh-coloured. The cock is quite black.

These birds are exceedingly voracious, and
feed on various species of fish and molluscæ.
The shoals of flying-fish, when persecuted by
their enemies of the deep, making their appearance for a short flight in the air, and suffer
greatly from the voracity of these birds. They
also often pursue the shoals of salmon into the
mouths of the large rivers; and so gorge them-

ordinary powers of flight, to be prevented by
their weight and consequent stupidity from even
rising.

In the West Indies the appearance of these
birds is said to foretel the arrival of ships; which
indeed is sometimes true, and arises from a very

natural cause. They always fish in fine weather;
so that when the wind is boisterous out at 'sea,
they retire into the harbours, where they are protected by the land: and the same wind that
blows them in, brings also very often vessels to
seek a retreat from the storm.

The voice of these birds resembles very much
the braying of the ass. In South America they
build their nests about the end of September;
these are formed of earth, on the ground, and
are from one to three feet high. The eggs are
as large as those of the goose, and have the singular property of their white not becoming hard
by boiling. When attempted to be seized, these
birds make a vigorous defence with their bills.

Many of the Indians set a high value on their
feathers; which they use for arrows, as they last
much longer than those of any other birds. The
natives of the South-Sea Islands watch the arrival of the man-of-war birds at the rainy season;
and, when they observe them, they launch from
their canoes a light float of wood into the water,
baited with a small fish. When one of the birds
approaches it, a man stands ready with a pole,
of about eighteen feet long; and on its pouncing,
he strikes at it, and seldom fails of bringing it
down. If, however, he misses his aim, he must
wait for some other bird; for that will no more
be tempted to approach. The cock birds are
reckoned the most valuable; and sometimes
even a large hog is given in exchange for one of
these.

The Alcatraci of the West Indies.

The inhabitants of Kamtschatka make buoys
to their nets, of the intestines of the man-of-war
birds, which they blow up like bladders. They
also make tobacco-pipes and needle cases, of the
bones of the wings; and use them too for heck-
ling the glass, which serves them instead of flax.
The flesh is very hard and dry.

This bird seems to be a species of the Alcatraci, only with this difference, the one frequents
the East, and the other the West Indies: he has
a great red comb, or rather wattle, hanging
under his throat like a cock's, which does not
appear in the young till they are full grown; the
females have none; there is no difficulty in taking of them when they sit on the ground, or in
their nest, on account of the length of their
wings, which hinders them from rising hastily,
so that they may be beaten down with long
sticks. The oil or fat of this bird is accounted a
sovereign remedy for several disorders proceeding

dies. It is a sea fowl, but its feet are not webbed
as the East India ones are.

A Portuguese traveller describes a bird he fre-

pira, otherwise the forked-tail, the tail being di-

says is good in particular cases, and the feathers
serve the Indians for their arrows: he adds, that
they are certain fore-runners of the arrival of the
ships, it very seldom failing, that some days after
they are seen, the ships arrive in the ports.

This, by the description he gives, seems to be the bird before described; but it is the misfortune of a great many travellers not only to mistake the different species, by some little variations that appear either in colour or size of the birds in climates a considerable distance from each other, but to give us their description in such terms, as are very often difficult to be understood.

THE SEA-LARK

IS a small bird that does not weigh more than two ounces, whose bill is short in comparison of other water-fowls, the upper part of which is encompassed with a black line, which spreads itself round the eyes, and passes across the middle of the head, where it encircles a fillet or broad bed of white, which runs from the inner corner of one eye to the other; the inner part of the throat is white, and round the neck there runs a fine collar, or double ring, the upper part of which is white, the lower part pretty broad and black, the rest of the body is of a dark ash-colour, except the breast and belly, which are white. The legs and feet are of an orange colour, the claws black, the tail about two inches long.

It builds its nest of grass, straw, and stalks of plants, upon the sea rocks where it lays greenish-coloured eggs, with brown spots upon them. It commonly makes but short flights, but runs exceedingly swift upon the shores, continuing to sing, and cry as it flies. It is common upon most of the sea coasts in England, and upon the banks of rivers, and is said to feed upon beetles and small insects.

The flesh is said by Aristotle, Ælian, Heleodorus, and others, to be a certain cure for the jaundice; and some authors have not scrupled to affirm, that the bare looking upon this bird is a present remedy for that distemper.

This, together with its property of hiding itself all day, and only flying abroad in the night, gave rise to the proverb—like a sea-lark, applied to such persons as enviously concealed any thing, the knowledge of which might be of public use or benefit.

CORMORANT, OR CORVORANT.

THE weight of this bird is about seven pounds,

assist it in holding its fishy prey. The head and

thick and heavy, more inclining in figure to that

till near the end, where the upper chap bends

into a hook; in the lower mandible there is a naked yellowish pouch; the tail is about five inches long, composed of hard stiff feathers, and the legs are strong and thick, but very short. The male has the feathers under the chin white, and likewise a short loose pendent crest; and part of the wings is sometimes of a deep and glossy blue green.

But notwithstanding the seeming heaviness of its make, there are few birds can exceed it in power of wing, or strength of flight. As soon as the winter approaches they are seen dispersed along the sea shore, and ascending up the mouths of fresh-water rivers, carrying destruction to all the finny tribe. They are most remarkably voracious, and have a most sudden digestion. Their appetite is for ever craving, and never satisfied. This gnawing sensation may probably be increased by the great quantity of small worms that fill their intestines, and which their unceasing gluttony contributes to engender.

Thus formed with the grossest appetites, this unclean bird has a very rank and disagreeable smell, and is more fœtid, than even carrion, in its most healthful state. Their flesh is so very disgusting, that even the Greenlanders (among whom they are very common) will scarcely eat them. "Its form," says the ingenious Mr. Pennant, "is disagreeable; its voice is hoarse and croaking; and all its qualities obscene. No wonder then that Milton should make Satan personate this bird, when he sent him upon the

basest purposes, to survey with pain the beauties
of life. It has been remarked, however, of our
poet, that the making a water-fowl perch on a
tree, implied no great acquaintance with the history of nature. In vindication of Milton, Aristotle expressly says, that the cormorant is the
only water-fowl that sits on trees. We have
already seen the pelican of this number; and the
cormorant's toes seem as fit for perching upon
trees as for swimming; so that our epic bard
seems to have been as deeply versed in natural
history as in criticism."

Bishop Newton, in his remarks on the following lines of Milton, also defends the poet's choice
of this voracious sea-fowl as a " proper emblem
of the destroyer of mankind."

> " Thence up he flew, and on the tree of life,
> The middle tree and highest there that grew,
> Sat like a cormorant; yet not true life
> Thereby regain'd, but sat devising death
> To them who liv'd; nor on the virtue thought
> Of that life-giving plant, but only us'd
> For prospect, what well us'd had been the pledge
> Of immortality."

" Indeed," says a modern writer, " this bird

fish are to be found, watches their migrations.
It is seen as well by land as sea; it fishes in fresh-
water lakes, as well as in the depths of the ocean;
it builds in the cliffs of rocks, as well as on trees;

Bred for the purpose of fishing.

Its indefatigable nature, and its great power in catching fish, were probably the motives that induced some nations to breed this bird up tame, for the purpose of fishing; and Willoughby assures us, it was once used in England for that purpose. The description of their manner of fishing is thus delivered by Faber. " When they carry them out of the rooms where they are kept, to the fish-pools, they hood-wink them, that they may not be frightened hy the way. When they are come to the rivers, they take off their hoods; and having tied a, leather thong round the lower part of their necks, that they may not swallow down the fish they catch, they throw them into the river. They presently dive under water; and there for a long time with wonderful swiftness, pursue the fish: and when they have caught them, instantly rise to the top of the water, and pressing the fish lightly with their bills, swallow them; till each bird hath after this manner, devoured five or six fishes. Then their keepers call them to the side, to which they readily fly; and, one after another, vomit up all their fish, a little bruised with the first nip given in catching them. When they have done fishing, setting the birds on some high place, they loose the string from their necks, leaving the passage to the stomach free and open; and for their reward, they throw them part of their prey; to each one or two fishes, which they will catch most dexterously, as they are falling in the air."

At present, the cormorant is trained up in every part of China for the same purpose, where there are many lakes and canals. " To this end," says Comte de Buffon, " they are educated as men rear up spaniels or hawks, and one man can easily manage an hundred. The fisher carries them out into the lake, perched on the gunnel of his boat, where they continue tranquil, and expecting his orders with patience. When arrived at the proper place, at the first signal given, each flies a different way to fulfil the task assigned it. It is very pleasant on this occasion, to behold with what sagacity they portion out the lake or the canal where they are upon duty. They hunt about, they plunge, they rise an hundred times to the surface, until they have at last found their prey. They then seize it with their beak by the middle, and carry it without fail to their master. When the fish is too large, they then give each other mutual assistance; one seizes it by the head, the other by the tail, and in this manner carry it to the boat together. There the boatman stretches out one of his long oars, on which they perch, and being delivered of their burthen, they fly off to pursue their sport. When they are wearied, he lets them rest for a while ; but they are never fed till their work is over. In this manner they supply a very plentiful table; but still their natural gluttony cannot be reclaimed even by education. They have always, while they fish, the same string fastened round their throats, to prevent them from devouring their prey, as otherwise they would at once satiate themselves, and discontinue the pursuit the moment they had filled their bellies."

The following account of this Chinese bird, by Sir George Staunton, is the most authentic of any that has yet been given to us.

" The embassy," he says, " had not proceeded far on the southern branch of the Imperial Canal, when they arrived in the vicinity of a place where the Leu-tze, or famed fishing bird of China, is bred and instructed in the art and practice of supplying his owner with fish in great abundance. It is a species of the pelican, resembling the common cormorant; but on a specimen being submitted to Dr. Shaw, he has distinguished it in the following terms.

" Brown pelican, or cormorant, with white throat, the body whitish beneath; the tail rounded; the irides blue; the bill yellow.

" On a large lake close to this part of the canal, and to the eastward of it are thousands of small boats and rafts built entirely for this species of fishing. On each boat, or raft, are ten or a dozen birds, which, at a signal from the owner, plunge into the water, and it is astonishing to see the enormous size of the fish with which they return, grasped within their bills. They

appear to be so well trained, that it did not require either ring or cord about their throats to
prevent them from swallowing any portion of
their prey, except what the master was pleased to

The boat used by these fishermen is of a remarkably light make, and is often carried to the lake,
together with the fishing-birds, by the men who
are there to be supported by it." See the annexed engraving.

The cormorant is the best fisher of all birds;
and though fat and heavy with the quantity it
devours, is nevertheless generally upon the wing.
The great activity with which it pursues, and

prey, offers one of the most amusing spectacles
to those who stand on the shore. This large
bird is seldom seen in the air, but where there
are fish below; and then they must be near the
surface, before it will venture to souse upon
them. If they are at a depth beyond what the
impetus of its flight makes the cormorant capable of diving to, they certainly escape him; for
this bird cannot move so fast under water, as the

unsuccessful dip; and is often seen rising heavily,
with a fish larger than it can devour. Goldsmith

has caught the fish by the tail; and, consequently, the fins prevent its being easily swallowed in that position. In this case, the bird is
seen to toss its prey above its head, and very

Remarkable voracity.

dexterously to catch it when descending, by the proper end, and so swallow it with ease."

Cormorants are common on many of our sea coasts: building their nests on the highest parts of the cliffs, that hang over the sea; and laying three or more pale green eggs, about the size of those of a goose. In winter they disperse along the shores, and visit the fresh waters, where they commit great depredations among the fish. They are remarkably voracious; having a most sudden digestion, promoted, perhaps, by the infinite quantity of small worms that fill their intestines. They are very wary, except when they have filled their stomach; when they become so stupid, that it is frequently an easy thing to take them in a net, or even by means of a noose thrown over their heads. We are informed by the Rev. Mr. Bingley, that in the year 1798, he saw one that had been seized by the hand, when perched at the top of a rock just behind the town of Caernarvon; and in the year 1793 one of them was observed sitting on the vane of St. Martin's steeple, Ludgate Hill, London, and was shot from thence in the presence of a great number of people.

It is no uncommon thing to see twenty of these birds together on the rocks of the sea coast, with extended wings, drying themselves in the wind; in this position they remain sometimes nearly an hour without once closing the wings; and as soon as these are sufficiently dry to enable the feathers to imbibe the oil, they press this substance from the receptacle on their rumps, and dress the feathers with it. It is only in one particular state that the oily matter can be spread on them; when they are somewhat damp; and the instinct of the birds teaches them the proper moment.

The skins of the cormorants are very tough; and are used by the Greenlanders, when sewed together and put into proper form, for garments. And the skin of the jaws, like that of others of this tribe, serves these people for bladders to buoy up their smaller kinds of fishing darts.

When cormorants were trained in this country, for the purpose of catching fish, they were kept with great care in the house: and on their being taken out for fishing, they had round their neck a leather thong, to prevent them from swallowing their prey: they were also hooded till brought to the water's edge. It appears that King Charles the First had an officer in his household, entitled " Master of the Corvorants."

The ambassador from the Duke of Holstein, in his travels into Muscovy and Persia, speaks of a kind of large wild geese, or cormorants, which they met with, and which the Muscovites call babbes. This author describes them as being larger than swans; and that their bills were above a foot long, two fingers broad, and forked at the end; under the bill, he says, they had a bag of skin, which they could contract quite close, or extend it to such a size as to contain more than

two gallons of liquor; and this they used as a re-
servatory for the fish they take, until they can
swallow them; he adds that one of them that was
shot upon the Caspean sea, measured two ells
and a half between the two extremities of the
wings, and seven feet from the head to the end
of the toes.   In this measurement, we may possibly not unfairly conclude, that some little allow-
ances must be made at his being struck with
astonishment at the animal himself, and wishing
to convey the same astonishment to others.   Fer-
nandez says, there are cormorants in Mexico,
which have teeth within their bills.

SHAG, OR LESSER CORMORANT.

THE common shag, which is another of the
cormorant genus, is in length two feet six inches,
and the extent of its wings eight feet.   The general colour of its plumage is black, the belly is
dusky, and the head and neck glossed with green.
The crested shag is somewhat less than the preceding, and is less common.   There are two kinds
which are natives of Kamtschatka; these are
distinguished as the violet and the red, faced
shags being so ornamented with those colours.
Besides there are several others found in New
Zealand, and also in Africa, in the latter of which
there are two species not larger than a teal.   The
whole of these like the cormorants build in trees.

THE RED-BACKED PELICAN.

THIS bird, like others of its race, is very voracious. Mr. *Lewis*, a navy surgeon, described
to Dr. Latham the mode in which some of them
that had been brought up tame, stowed its food
into its pouch. " A number of different sized
fishes were laid before it on the ground; it first
attempted to take up one that weighed ten
pounds, but the bill was much too weak for this
exertion; it however, picked up as many as ten
others, each of which weighed about a pound,
and arranged them in rows with their heads towards the throat: and after this, it walked off in
a very stately manner, with the bag hanging down
to its feet. The pouch held two gallons of water."

THE GREAT GREY GULL

WEIGHS twelve or fourteen ounces, and is
from the point of the bill to the extremity of the
tail, about twenty inches; and from the point of

The bill is black, and near three inches long, the
upper mandible something longer than the under,
and a little hooked, or inclining downwards over
it; the lower has a rising, or crooked set, towards
the extremity: the eyes are grey, the nostrils in
a sort of oblong form, the head very large; the
necks of these birds in general are so short, that
when they walk or stand, they appear so much

sunk, or drawn towards the shoulders, as one would almost imagine they had not any neck at all.

The upper side of the back and neck, are grey, intermixed with a whitish brown; the back feathers black in the middle, and ash-coloured towards the edges, the wing feathers are of a dark brown, intermixed with black; the throat, breast, belly, and thighs, white; the rump is of the same colour, with a few brown spots interspersed. The tail is five or six inches long, the outmost tip of the feathers, on the upper side, are joined by a sort of black cross bars, near two inches broad; the under part also varied with a few dusky-coloured lines. The legs and feet are yellow-orange coloured, and the claws black.

There are about twenty varieties of this tribe which are all distinguished by an angular knob on the chap. This bird is well known. " It is," says an ingenious modern, " seen with a slow-sailing flight hovering over rivers to prey upon the smaller kinds of fish; it is seen following the ploughman in fallow fields to pick up insects; and when living animal food does not offer, it has even been known to eat carrion, and whatever else of the kind that it finds. Gulls are found in great plenty in every place; but it is chiefly round our rockiest shores that they are seen in the greatest abundance; it is there that the gull breeds and brings up its young; it is there that millions of them are heard screaming with discordant notes for months together.

" Those who have been much upon our coasts know that there are two different kinds of shores; that which slants down to the water with a gentle declivity, and that which rises with a precipitate boldness, and seems set as a bulwark to repel the force of the invading deeps. It is to such shores as these that the whole tribe of the gull-kind resort, as the rocks offer them a retreat for their young, and the sea a sufficient supply. It is in the cavities of these rocks, of which the shore is composed, that the vast variety of seafowls retire to breed in safety. The waves beneath, that continually beat at the base, often wear the shore into an impending boldness; so that it seems to jut out over the water, while the raging of the sea makes the place inaccessible from below. These are the situations to which sea-fowl chiefly resort, and bring up their young in undisturbed security.

" Those who have never observed our boldest coasts have no idea of their tremendous sublimity. The boasted works of art, the highest towers, and the noblest domes, are but ant-hills when put in comparison: the single cavity of a rock often exhibits a coping higher than the cieling of a gothic cathedral. The face of the shore offers to the view a wall of massive stone; ten times higher than our tallest steeples. What should we think of a precipice three quarters of a mile in height? And yet the rocks of St. Kilda are still higher! What must be our awe to approach the edge of that impending height, and

to look down on the unfathomable vacuity below; to ponder on the terrors of falling to the
bottom, where the waves that swell like mountains are scarcely seen to curl on the surface, and
the roar of a thousand leagues broad appears
softer than the murmurs of a brook? It is in
these formidable mansions that myriads of seafowls are for ever seen sporting, flying in security down the depth, half a mile beneath the feet
of the spectator. The crow and the chough
avoid those frightful precipices; they chuse
smaller heights, where they are less exposed to
the tempest; it is the cormorant, the gannet, the
tarrock, and the terne, that venture to those
dreadful retreats, and claim an undisturbed possession. To the spectator from above, those birds,
though some of them are above the size of an
eagle, seem scarce as large as a swallow; and
their loudest screaming is scarcely audible.

" But the generality of our shores are not so
formidable. Though they may rise two hundred
fathoms above the surface, yet it often happens
that the water forsakes the shore at the departure of the tide, and leaves a noble and delightful walk for curiosity on the beach. Not to men-
tion the variety of shells with which the sand
is strewed, the lofty rocks that hang over the
spectator's head, and that seem but just kept
from falling, produce in him no unpleasing
gloom. If to this be added the fluttering, the
screaming, and the pursuits of myriads of water-

birds, all either intent on the duties of incubation, or roused at the presence of a stranger, nothing can compose a scene of more peculiar solemnity. To walk along the shore when the tide is departed, or to sit in the hollow of a rock when it is come in, attentive to the various sounds that gather on every side, above and below, may raise the mind to its highest and noblest exertions. The solemn roar of the waves swelling into and subsiding from the vast caverns beneath, the piercing note of the gull, the frequent chatter of the guillemot, the loud note of the auk, the scream of the heron, and the hoarse deep periodical croaking of the cormorant, all unite to furnish out the grandeur of the scene, and turn the mind to HIM who is the essence of all sublimity.

" Yet it often happens that the contemplation of a sea-shore produces ideas of an humbler kind, yet still not unpleasing. The various arts of these birds to seize their prey, and sometimes to elude their pursuers, their society among each other, and their tenderness and care of their young, produce gentle sensations. It is ridiculous also now and then to see their various ways of imposing upon each other. It is common enough, for instance, with the arctic gull, to pursue the lesser gulls so long, that they drop their excrements through fear, which the hungry hunter soon gobbles up before it ever reaches the water. In breeding too they have frequent

Occasional battles.

contests: one bird who has no nest of her own,
attempts to dispossess another, and put herself
in the place. This often happens among all the
gull-kind; and the bird, thus displaced by her
more powerful invader, will sit near the nest in
pensive discontent, while the other seems quite
comfortable in her new habitation. Yet this
place of pre-eminence is not easily obtained; for
the instant the invader goes to snatch a momen-
tary sustenance, the other enters upon her own,
and always ventures another battle before she re-
linquishes the justness of her claim. The contemplation of a cliff thus covered with hatching-
birds, affords a very agreeable entertainment;
and as they sit upon the ledges of the rocks,
one above another, with their white breasts forward, the whole group has not unaptly been
compared to an apothecary's shop.

" These birds, like all others of the rapacious
kind, lay but few eggs; and hence, in many
places, their number is daily seen to diminish.
The lessening of so many rapacious birds may,
at first sight, appear a benefit to mankind; but
when we consider how many of the natives of
our islands are sustained by their flesh, either
fresh or salted, we shall find no satisfaction in.
thinking that these poor people may in time lose
their chief support. The gull, in general, as was
said, builds on the ledges of rocks, and lays from
one egg to three, in a nest formed of long grass
and sea-weed. Most of the kind are fishy tasted,

with black stringy flesh; yet the young ones are better food; and of these, with several other birds of the penguin kind, the poor inhabitants of our northern islands make their wretched banquets. They have been long used to no other food; and even a salted gull can be relished by those who know no better."

THE BROWN GULL

IS considerably less than the former, the bill is about an inch and a half long, black towards the extremity, the rest of a light brown or horn colour, shaped much like the former; the eyes are small, the circles yellow, and the nostrils in an oblong form. The head and all the upper parts of the body and wings are of a dusky sort of brown colour, except some of the prime feathers of the wings, which are quite black. The belly and breast are of a more bright colour, interspersed with a considerable number of transverse brown lines. The tail is black, the legs and feet of a brownish yellow, and the claws black.

This seems to be an uncommon bird, and not known to authors that have written upon the subject, being classed among the gull-kind, chiefly from the resemblance of its bill and legs. Mr. Albin says, it seems to be a non-descript bird.

THE BROWN-HEADED GULL

IS much about the size of the preceding; the bill is red and sharp pointed; the under mandible bunching out into a small angle, the eyes black, the irides, or circles, red; encompassed with a broad circle of pale, or white feathers; the head and neck brown, the lower part towards the breast more dusky; the covert feathers of the wings and the back are of an ash-colour, the prime feathers black, with their outer edges, or webs white; the rest of the body white, tinctured with a yellowish sort of pale green. The tail is near five inches long, the legs and feet red, and the claws black.

These birds are common about Gravesend, in the river Thames.

THE BLACK AND WHITE GULL

IS by far the largest of all the gull-kind, weighing generally upwards of four pounds, and being twenty-five or twenty-six inches, from the point of the bill to the end of the tail; and from the tip of each wing, when extended, five feet and several inches. The bill appears compressed sideways, being more than three inches long, and hooked towards the end, like the rest of this

5

kind, of a sort of orange colour; the nostrils in an oblong form; the mouth wide,, with a long tongue and very open gullet.

The irides of the eyes are of a very delightful red. The wings, and the middle of the back are black, only the tips of the covert and quill feathers are white. The head, breast, tail, and other parts of the body are likewise white. The tail is near six inches long, the legs and feet flesh-coloured, and the claws black. It is a sea-fowl, and preys upon fishes, which have been taken whole from its stomach.

THE WHITE GULL.

THIS is one of the smallest sort, and does not weigh above eight or nine ounces; the form of the bill is very much like those before described, and of a red colour, with an angle on the lower mandible: the irides of the eyes white, encircled with an ash colour.

The prime feathers on the wings are black, the

beyond the tail; the back and covert feathers grey, or ash-coloured; the head, breast, throat, and belly white, tinctured with a pale or faint yellow. The legs are bare of feathers above the knees, and of a dusky green colour, the claws small, but more dusky and blackish.

Fierce in defence of its young.

They are said to be useful in gardens, where they destroy the insects and worms; their food is chiefly small fish.

The birds of this kind, are in many places called sea-mews, in others sea-cobs.

THE SKUA GULL.

THIS species is nearly two feet in length, and weighs about three pounds. Its bill is two inches and a quarter long, hooked at the end, and very sharp; and the upper mandible is covered more than halfway down, with a black cere or skin, as in the hawk kind. The feathers of the upper parts are of a deep brown, but below they are somewhat of a rust colour. The talons are black, strong, and crooked.

The skua gull inhabits Norway, the Feroe Islands, Shetland, and the noted rock Foula, a little west of these last. It is the most formidable of the tribe; its prey being not only fish, but (what is wonderful, in a web-footed bird) all the lesser sorts of water-fowl, and (according to the account of Mr. Schroter, a surgeon of the Feroe Isles) ducks, poultry, and even young lambs.

This bird has all the fierceness of the eagle in defending its young. When the inhabitants of those islands visit the nest, it attacks them with such force, that, if they hold a knife perpendicularly over their heads, the gull will sometimes transfix itself in its fall on the plunderers. '1 Rev. Mr. Low, minister of Birfa, in Orkney, forms us, that on his approaching the habitati of these birds, they assailed him, and the cc pany along with him, in the most violent mann and intimidated a bold dog in such a manner to drive him for protection to his master. '] natives are often very rudely treated by th while they are attending their cattle on the hi and are frequently obliged to guard their he by holding up their sticks, on which (in manner mentioned above) the birds often themselves.

In Foula the skua gulls are privileged: be said to defend the flocks from the attacks of eagle, which they beat off and pursue with gr fury; so that even that rapacious bird seldom v tures to approach the places where they inha The natives of Foula on this account impose a 1 upon any person who destroys one of these t ful defenders: and deny that they ever inj their flocks or poultry; but imagine them to l only on the dung of the Arctic gull and ot larger birds.

The following is the account, given in Jac son's History of the Feroe Islands, of the metl in which these birds are taken: " It cannot expressed with what pains and danger they t these birds in those high steep cliffs, whei many are two hundred fathoms high. But th are men apt by nature and fit for the work, v

take them usually in two manners: they either
climb from below into these high promontories,
that are as steep as a wall; or they let themselves
down with a rope from above. When they
climb from below, they have a pole five or six
ells long, with an iron hook at the end, which
they that are below in the boat, or on the cliff,
fasten unto the man's girdle, helping him up thus
to the highest place where he can get footing:
afterwards they also help up another man; and
thus several climb up as high as possibly they
can; and where they find difficulty, they help
each other up by thrusting one another up with
their poles. When the first hath taken footing,
he draws the other up to him, by the rope fastened to his waist; and so they proceed, till they
come to the place where the birds build. They
there go about as well as they can, in those dangerous places; the one holding the rope at one
end, and fixing himself to the rocks; the other
going at the other end from place to place. If
it should happen that he chanceth to fall, the
other that stands firm keeps him and helps him
up again. But if he passeth safe, he likewise
fastens himself till the other has passed the same
dangerous place also. Thus they go about the
cliffs after birds as they please. It often hap-
peneth, however, the more is the pity, that
where one doth not stand fast enough, or is not
sufficiently strong to hold up the other in his fall,

that they both fall down and are killed.   In this manner some do fall every year."

Mr. Peter Clauson, in his Description of Norway, states, that there was anciently a law in that country, that whosoever climbed so on the cliffs, that he fell down and died, if the body was found, before burial, his next kinsman should go the same way ; but if he durst not, or could not do it, the dead body was not then to be buried in sanctified earth, as the person was too full of temerity, and his own destroyer.

" When the fowlers," continues Jacobson, " get, in the manner aforesaid, to the birds within the cliffs, where people seldom come, the birds are so tame that they take them with their hands; for they will not leave their young.   But when they are wild they cast a net, with which they are provided, over them, and entangle them therein.   In the mean time, there lieth a boat beneath in the sea, wherein they cast the birds killed: and in this manner they can, in a short time, fill a boat with fowl.   When it is pretty fair weather, and there is good fowling, the fowlers stay in the cliffs seven or eight days together; for there are here and there holes in the rocks, where they can safely rest; and they have meat let down to them with a line from the top of the mountain.   In the mean time some go every day to them, to fetch home what they have taken.

" Some rocks are so difficult, that they can in no manner get unto them from below; wherefore they seek to come down thereunto from above. For this purpose they have a rope, eighty or a hundred fathoms long, made of hemp, and three inches thick. The fowler maketh the end of this fast about his waist, and between his legs, so that he can sit thereon; and is thus let down, with the fowling staff in his hand. Six men hold by the rope, and let him easily down, laying a large piece of wood on the brink of the rock, upon which the rope glideth, that it may not be worn to pieces by the hard and rough edge of the stone. They have besides, another small line that is fastened to the fowler's body; on which he pulleth, to give them notice how they should let down the great rope, either lower or higher; or to hold still, that he may stay in the place whereunto he is come. Here the man is in great danger, because of the stones that are loosened from the cliff, by the swinging of the rope, and he cannot avoid them. To remedy this, in some measure, he hath usually on his head a seaman's thick and shaggy cap, which defends him from the blows of the stones, if they be not too big; but if they are, which is frequently the case, it costeth him his life: nevertheless, they continually put themselves in that danger for the wretched body's food sake, hoping in God's mercy and protection, unto which the greatest part of them devoutly recommend

themselves when they go to work: otherwise,
they say, there is no other great danger in it,
except that it is a toilsome and artificial labour; for he that hath not learned to be so let
down, and is not used thereto, is turned about
with the rope, so that he soon groweth giddy,
and can do nothing; but he that hath learned
the art, considers it as a sport, swings himself on
the rope, sets his feet against the rock, casts himself some fathoms from thence, and shoots himself to what place he will; he knows where the
birds are, he understands how to sit on the line
in the air, and how to hold the fowling-staff in
his hand; striking therewith the birds that come
or fly away: and when there are holes in the
rocks, and it stretches itself out, making underneath as a ceiling, under which the birds are, he
knoweth how to shoot himself in among them,
and there take firm footing. There, when he is
in these holes, he maketh himself loose of the
rope, which he fastens to the crag of the rock,
that it may not slip from him to the outside of
the cliff. He then goes about in the rock, taking the fowl, either with his hands or with the
fowling-staff. Thus, when he hath killed as
many birds as he thinks fit, he ties them in a
bundle, and fastens them to a little rope, giving
a sign, by pulling, that they should draw them
up. When he has wrought thus the whole day,
and desires to get up again, he sitteth once more
upon the great rope, giving a new sign that they

should pull him up; or else he worketh himself
up, climbing along the rope, with his girdle full
of birds. ᵎ It is also usual, where there are not
folks enough to hold the great rope, for the
fowler to drive a post sloping into the earth, and
to make a rope fast thereto, by which he lets
himself down without any body's help, to work
in the manner aforesaid. Some rocks are so
formed that the persons can go into their cavities
by land.

"These manners are more terrible and dangerous to see than to describe; especially if
one considers the steepness and height of the
rocks, it seeming impossible for a man to approach them, much less to climb or descend. In
some places, the fowlers are seen climbing where
they can only fasten the ends of their toes and
fingers; not shunning such places, though there
be an hundred fathom between them and the
sea. It is dear meat for these poor people, for
which they must venture their lives; and many,
after long venturing, do at last perish therein.

"When the fowl is brought home, a part
thereof is eaten fresh; another part, when there
is much taken, being hung up for winter provision. The feathers are gathered to make mer-
chandize of, for other expences. The inhabitants get a great many of these fowls, as God
giveth his blessing and fit weather. When it is
dark and hazy they take the most; for then the
birds stay in the rocks: but in clear weather,

2

and hot sun-shine they seek the sea. When
they prepare to depart for the season, they keep
themselves most there, sitting on the cliffs towards the sea-side, where the people get at them
sometimes with boats, and take them with fowl-
ing-staves."

Strange and almost incredible as the above account may appear, the circumstances are too
well known to leave the smallest doubt of this
author's veracity; and the hardihood of the people who inhabit the rocky shores of the northern
parts of Europe, in these pursuits, is almost pro-
verbial; with many of them the birds so taken
constitute the chief part of their food, and therefore, possibly, it is necessity has taught them to
put danger at defiance.

THE HERRING GULL, AND OTHERS.

THE herring gull resembles the black and
white gull in every thing but size, except that the
plumage on the back and wings is more inclined
to ash-colour than black; it weighs thirty ounces.
The glacous gull, which inhabits Norway, &c. is
rather larger than the herring gull, but resembles
it in most other respects. The silvery gull is the
same size as the herring gull, and not much different in plumage and manners.

The tarrock, and kittiwake gulls, also so nearly
resemble each other, that some authors affirm

the latter to be only the tarrock in a state of perfection. The head, neck, belly, and tail of the kittiwake are of a snowy whiteness, the back and wings are grey; and both also have behind each ear a dark spot; both species are about the same size, viz. fourteen inches, and the tarrock weighs seven ounces. Of the arctic gull the male has the top of the head black; the back, wings, and tail dusky; the rest of the body white: the female is entirely brown.

. The pewit-gull, or black-cap, is so called from the head and throat being of a dark or black colour. The red-legged gull, the brown-throated gull, and the laughing gull, which only differs from the others in having the legs black instead of red, are possibly only varieties of the same species. They are in length from fifteen to eightteen inches. The back and wings of these birds are in general ash-coloured, and the rest of the body white. The young birds of these species are thought by some to be good eating.

The gnat gull, which is found on the borders of the Caspian Sea, though distinguished by a black head, is quite a different species from our black-cap. It is about the size of a barnacle-goose, and weighs between two and three pounds. Its voice is as hoarse as that of a raven.

THE PETREL.

THESE birds all frequent the ocean and are seldom to be seen on shore except during the breeding season. Their legs are bare of feathers a little above the knee. They have a singular faculty of spouting from their bills, to a considerable distance, a large quantity of pure oil; which they do, by way of defence, into the face of any one that attempts to annoy them. This oil has been frequently used in medicine; and some writers, say with success.

The bill is somewhat compressed; the mandibles are equal, and the upper one is hooked at the point. The nostrils form a truncated cylinder, lying over the base of the bill. The feet are webbed; and, in the place of a hind toe, have a spur, or nail, pointing downwards.

THE STORMY PETREL.

THE stormy petrel is not larger than a swallow; and its colour is entirely black; except the coverts of the tail, the tail itself, and the vent feathers, which are white. Its legs are long and slender.

It is found in most seas, and frequently at a vast distance from the land, where it braves the utmost fury of the storm, sometimes skimming

with incredible velocity along the hollows of the
waves, and sometimes over their summits. It is
also an excellent diver, and often follows vessels
in great flocks, to pick up any thing that is
thrown overboard; but its appearance is always
looked upon by the sailors as the sure presage of
stormy weather in the course of a few hours. It
seems to seek for protection from the fury of the
wind, in the wake of the vessels: and for the
same reason it very probably is, that it often
flies along between two surges.

The nests of these birds are found in the Orkney Islands, under loose stones, in the months of
June and July. They live chiefly on small fish;
and although mute by day, are very clamorous
during the night.

These birds are called by the sailors Mother
Carey's chickens; but why they have given
them this appellation we are at a loss to explain.
They are found in many parts of the world; and,
in the Feroe islands, the inhabitants are said to
draw a wick through the body of the bird, from
the mouth to the vent, which being lighted at
one end serves them as a candle, as it is fed by
the vast proportion of oil which this little animal
contains.

There are about twenty species of foreign birds
of this kind. In the high southern latitudes one
is found, which is the size of a goose, and on
that account is called the giant petrel. The

upper parts of its plumage are pale brown, mottled with dusky white; the under parts are white.

Mr. Anderson, in Capt. Cook's last voyage, mentions a petrel found at Kerguelan's Land, which the seamen called Mother Carey's Goose; it is by far the largest known; " they were," says he, " so tame, that at first we could kill them with a stick upon the beach. They are not inferior in size to an albatross, and are carnivorous, feeding on the carcasses of seals or birds, that were thrown into the sea. Their colour is a sooty brown, with a greenish bill and feet." This Mr. Anderson considered to be the same bird that is described by Pernetty, in his voyage to the Falkland Islands, and which is called *quebrantehuessos* by the Spaniards.

GREAT TERN, OR SEA SWALLOW,

IS about fourteen inches long, and weighs four ounces and a quarter. The bill and feet are a fine crimson, the former is tipped with black, and very slender. The back of the head is

and the under part white. The birds have been

the same actions at sea that the swallow has at

surface, and darting down upon the smaller fishes, which they seize with incredible rapidity.

The lesser tern weighs only two ounces ·five grains. The bill is yellow, and from the eyes to the bill is a black line. In other respects it almost exactly resembles the preceding.

THE BLACK TERN.

THIS is of a middle size between the two preceding species. It weighs two ounces and a half.' It receives its name from being all black as far as the vent, except a spot white under the throat. This bird is called in some parts the ear swallow. It is a very noisy animal. ·

Among the ·foreign birds ·of the. tern genus, there are some found of a snowy white; but the most singular bird of the kind is the striated tern, which is found at New Zealand. It is thirteen inches in length. The· bill is black, and the body in general mottled, or rather striped' with black and· white.' The noddy is about fifteen inches long, and the whole plumage a sooty brown, except the top of the head, which is white. It is a very common bird in the tropical seas, where it is known frequently to fly on board ships, and is taken with the hand. But though it be thus stupid, it bites the fingers severely, so as to make it unsafe to hold it. It is said to breed in the Bahama islands. · ·

THE FULMAR

IS the largest of the petrel kind which is known in these climates. It is superior to the size of the common gull, being about fifteen inches in length, and in weight seventeen ounces. The bill is very strong, yellow, and hooked at the end. The head, neck, and all the under parts of the body, are white; the back and wings ash-coloured, the quills dusky, and the tail white. It feeds on the blubber of whales, which supplies the reservoir, whence it spouts with a constant stock of ammunition. This oil is esteemed by the inhabitants of the North as a sovereign remedy in many complaints both external and internal. The flesh is also considered by them as a delicacy, and the bird is therefore in great quest at St. Kilda. It is said that when a whale is taken, these birds will, in defiance of all endeavours, light upon it, and pick out large lumps of fat even while it is alive.

THE SHEARWATER.

THIS is something smaller than the preceding. The head, and all the upper parts of the body, are of a sooty blackness; and the under part and inner coverts of the wings white. These birds are found in the Calf of Man, and the

2

Scilly Isles. In February they take possession
of the rabbit furrows, and then disappear till.
April; they lay one egg, and in a short time the
young are fit to be taken. They are then salted
and barrelled. During the day they keep at
sea fishing, and towards evening return to their
young, whom they feed by discharging the con-
tents of the stomach into their mouths.

THE GANNET, OR SOLAND GOOSE.

THIS bird, which is about the size of a tame
goose, is somewhat more than three feet in
length, and weighs about seven pounds. The
bill is six inches long: straight almost to the
point, where it is a little bent; its edges are irre-
gularly jagged, for the better securing of its prey;
and about an inch from the base of the upper
mandible, is a sharp process, pointing forward.
The general colour of the plumage is dirty white,
with a cinereous tinge. Surrounding each eye
there is a naked skin of fine blue: from the corner of the mouth a narrow slip of naked black
skin extends to the hind part of the head; and
beneath the chin is a pouch, like that of the pelican, capable of containing five or six herrings.
The neck is long; the body flat, and very full of
feathers. On the crown of the head, and the
back part of the neck, is a small buff-coloured
space · The quill feathers, and some other parts

of the wings, are black; as are also the legs, ex
cept a fine pea-green stripe in their front. The
tail is wedge-shaped, and consists of twelve sharp-
pointed feathers.

These birds, which subsist entirely upon fish,
chiefly resort to those uninhabited islands where
their food is found in plenty, and men seldom
come to disturb them. The islands to the north
of Scotland, the Skelig Islands of the coasts of
Kerry, in Ireland, and those that lie in the north
sea off Norway, abound with them. But it is on
the Bass Island, in the Firth of Edinburgh,
where they are seen in the greatest abundance.
" There is a small island," says the celebrated
Harvey, " called the Bass, not more than a mile
in circumference. The surface is almost wholly
covered during the months of May and June
with their nests, their eggs, and young. It is
scarcely possible to walk without treading on
them; the flocks of birds upon the wing are so
numerous, as to darken the air like a cloud; and
their noise is such, that one cannot, without difficulty, be heard by the person next to him.
When one looks down upon the sea from the
precipice, its whole surface seems covered with
infinite numbers of birds of different kinds, swimming and pursuing their prey. If, in sailing
round the island, one surveys its hanging cliffs,
in every crag or fissure of the broken rocks, may
be seen innumerable birds, of various sorts and
sizes, more than the stars of heaven, when viewed

in a serene night. If they are viewed at a distance, either receding, or in their approach to the island, they seem like one vast swarm of bees."

They are not less frequent upon the rocks of St. Kilda. Martin assures us, that the inhabitants of that small island consume annually near twenty-three thousand young birds of this species, besides an amazing quantity of their eggs. On these they principally subsist throughout the year; and from the number of these visitants, make an estimate of their plenty for the season. They preserve both the eggs and fowls in small pyramidal stone buildings, covering them with turf-ashes, to prevent the evaporation of their moisture. These birds are very voracious, yet somewhat dainty in their choice of prey, disdaining to eat any thing worse than herrings or mackerel, unless in great want. Allowing that these birds remain at St. Kilda about six months in the year, and that each bird destroys five herrings in a day (which is considerably less than the average) we have at least ninety millions of these, the finest fish in the world, devoured annually by a single species of St. Kilda birds.

The gannet, or Soland goose, is a bird of passage. In winter it seeks the more southern coasts of Cornwall, hovering over the shoals of herrings and pilchards that then come down from the northern sea: its first appearance in the northern islands is in the beginning of spring; and it continues to breed till the end of summer. But, in general, its motions are determined by the migrations of the immense shoal of herrings that come pouring down at that season through the British Channel, and supply all Europe as well as this bird with their spoil. The gannet assiduously attends the shoal in their passage, keeps with them in their whole circuit round our island, and shares with our fishermen this exhaustless banquet. As it is strong of wing, it never comes near the land, but is constant to its prey. Whereever the gannet is seen, it is sure to announce to the fishermen the arrival of the finny tribe; they then prepare their nets, and take the herrings by millions at a draught; while the gannet, who came to give the first information, comes, though an unbidden guest, and snatches its prey from the fisherman even in his boat. While the fishing season continues, the gannets are busily employed: but when the pilchards disappear from our coasts, the gannet takes its leave to keep them company.

The cormorant has been remarked for the quickness of his sight; yet in this the gannet seems to exceed him. It is possessed of a transparent membrane under the eye-lid, with which it covers the whole eye at pleasure, without ob-

seems a necessary provision for the security of the eyes of so weighty a creature, whose method of taking prey, like that of the cormorant, is by darting headlong down from an height of an hundred feet and more into the water to seize it. These birds are sometimes taken at sea, by fastening a pilchard to a board, which they leave floating. The gannet instantly bounces down from above upon the board, and is killed or maimed by the shock of a body where it expected no resistance.

The gannets breed but once a year, and lay only one egg, but if that be taken away, they lay another; and if that be also taken away, then a third; but never more for that season. Their eggs are white, and rather less than those of the common goose; and their nest large, composed of such substances as are found floating on the surface of the sea. The young birds, during the first year, differ greatly in colour from the old ones; being of a dusky hue, speckled with numerous triangular white spots; and at that time resembling the colours of the speckled diver.

They come yearly to the Bass Island, which is an almost inaccessible rock, situated at the mouth of the Forth in Scotland, seven miles from land, and faces St. Andrews on the north, North Berwick on the south, and the German ocean on the east. It was anciently a kind of prison for those who dissented from the then established church. There they breed in great numbers; it belongs to one proprietor, and care is taken never to frighten away the birds when laying, or to shoot them upon the wing. By that means they be-

come so confident as to alight and feed their
young ones unconcerned at any person's being
near them. They feed upon fish, as we have
observed; yet the young gannet is counted a
great dainty by the Scots, and sold very dear;
so that the lord of the above islet makes a considerable annual profit by the quantity that is taken
therefrom.

They quit this island towards the latter end of
autumn, and when they return in the spring
there is usually but three or four at first, which
precede the rest as so many spies, or harbingers,
and are followed by the flock a few days after, as
is attested by several creditable authors. They
build their nests in the highest and steepest rocks
they can find near the sea, and employ for that
purpose such a quantity of sticks as is almost incredible; insomuch that the inhabitants of that
part of the country, upon finding a few nests,
think themselves plentifully provided with fur
for a twelvemonth. They deposit their eggs
in the holes of the rock, and while they are laying them, rest one foot upon another; whence
Johnson thinks they derive their name from So-
lea, the sole of the foot: but this is rather an improbable derivation. They feed their young ones
with the most delicate sort of fish; and if, in flying away with one, they see another they like
better, they immediately drop the first, and
plunge into the water again with great violence.

They likewise disgorge a great quantity of fish,
which were formerly used as food by the garrison
of the castle.   The young, during the first year,
differ greatly from the old ones; being of a
dusky hue, and speckled with numerous triangular white spots.   While the female is employed
in incubation, the male supplies her with food;
and the young itself extracts its food from the
pouch of the parent, with its bill as with a
pincer.

These birds, when they pass from place to
place, unite in small flocks of from five to fifteen: and, except in very fine weather, fly low,
near the shore, but never pass over it; doubling
the capes and projecting parts, and keeping
nearly at an equal distance from the land.   During their fishing they rise high into the air, and
sail aloft over the shoals of herrings or pilchards,
much in the manner of kites.   When they observe the shoal crowded thick together, they
close their wings to their sides and precipitate
themselves, head-foremost, into the water, dropping almost like a stone.   Their eye in this act
is so correct, that they never fail to rise with a
fish in their mouth.

Mr. Pennant says, that the natives of Saint
Kilda hold this bird in much estimation, and
often undergo the greatest risks to obtain it.
Where this is possible, they climb up the rocks
which it frequents; and in doing this they pass

pearance, to allow them barely room to cling, and that too at an amazing height over a raging sea. Where this cannot be done, the fowler is lowered by a rope from the top; and to take the young, often stations himself on the most dangerous ledges: unterrified, however, he ransacks all the nests within his reach; and then, by

places to do the same. (See the annexed engraving.) It has been also said by travellers, that to take the old birds, the inhabitants tie a herring to a board, and set it afloat; so that by falling furiously upon it, the bird may break its neck in the attempt. This, however is unlawful; for the fastening of herrings thus to planks at sea, to catch the Soland goose, or gannet, is forbidden under a severe penalty.

Some years ago one of these birds was flying

pilchards lying on a fir plank, in a place for curing

inch-and-a-quarter plank, and kill itself on the spot.

pilchards during their whole progress round the British Islands; and sometimes migrates in quest of food as far southward as the mouth of the Tagus, being frequently seen off Lisbon during the month of September. From this time till

TAKING SEA FOWL, &c.

March it is not well known what becomes of these birds.

The young birds, and the eggs alone are eatable; the old ones being tough and rancid.

THE BOOBY.

THIS bird is about two feet six inches in length. Its bill is nearly four inches and a half long; toothed on the edges, and of a grey colour. A space round the eyes, and on the chin, is naked. The head, neck, upper parts of the body, wings, and tail, are ash-coloured brown; and the breast, under parts, and thighs, white. The eggs are pale yellow, and the claws grey.

This and some other species have been denominated boobies from their excessive stupidity; their silly aspect; and their habit of continually shaking their head and shivering when they alight on the ship's yards, or other parts, where they often suffer themselves to be taken by the hand. In their shape and organization they greatly resemble the cormorants.

The boobies have an enemy of their own tribe, that perpetually harasses them. This is the frigate pelican, or man-of-war bird, which rushes upon them, pursues them without intermission, and obliges them, by blows with its wings and bill, to surrender the prey that they have taken, which it instantly seizes and swallows. Catesby thus describes the skirmishes of the booby and its enemy, which he calls the pirate. " The latter," he says, " subsists entirely on the spoils of others, and particularly of the booby. As soon as the pirate perceives that it has caught a fish, he flies furiously against it, and obliges it to dive under water for safety; the pirate not being able to follow it, hovers above the water till the booby is obliged to emerge for respiration, and then attacks it again while spent and breathless, and compels it to surrender its fish : it now returns to its labours, and has to suffer fresh attacks from its enemy."

Leguat says, the boobies repair at night to repose on the island of Rodrigue; and the frigate waits for them on the tops of the trees; it rises very high, and darts down upon them like a hawk upon his prey, not to kill them, but to make them disgorge. The booby, struck in this way by the frigate, throws up a fish, which the latter snatches in the air : often the booby screams, and discovers a reluctance to part with its booty; but the frigate scorns its cries, and rising again, descends with such a blow as to stun the poor bird, and compel an immediate surrender.

Dampier gives us a curious account of the hostilities between the man-of-war birds, and the boobies, in the Alcrane Islands, on the coast of

Yucatan. " These birds were crowded so thick, that I could not," he says, " pass their haunt without being incommoded by their pecking.— I observed that they were ranged in pairs, which made me presume that they were male and female. When I struck them, some flew away; but the greater number remained, and would not stir, notwithstanding all I could do to rouse them. I remarked also, that the man-of-war birds and the boobies always placed sentinels over their young, especially when they went to sea for provisions. Of the man-of-war birds, many were sick or maimed, and seemed unfit to procure their subsistence. They lived not with the rest of their kind; either expelled from society, or separated by choice: but were dispersed in different places, probably that they might have a better opportunity of pillaging. I once saw more than twenty on one of the islands, sally out from time to time into the open country to carry off booty, and return again almost immediately. When one surprised a young booby that had no guard, he gave it a violent peck on the back to make it disgorge, which it did instantly: it cast up one or two fish about the bulk of one's hand, which the old man-of-war bird swallowed still more hastily. The vigorous ones play the same game with the old boobies which they find at sea. I saw one myself which flew right against a booby; and with one stroke of its bill, made him deliver up a fish that he had just swallowed. The man-of-war bird darted so rapidly, as to catch this fish in the air before it could fall into the water."

THE CURLEW.

THERE are eleven species of this bird, according to Latham, differing very much in size, the longest measuring about twenty-five inches, and sometimes weighing thirty-six ounces. These birds fly in considerable flocks, and are well known upon the sea-coasts in most parts, where, and in the marshes, they frequent in the winter, feeding on worms, frogs, and all kinds of marine insects. In April, or the beginning of May, they retire into the mountainous and unfrequented parts of the sea shore, where they breed, and do not return again till the approach of winter. There have been some advocates in favour of the flesh of this bird, but in general it is strong, rank, and fishy. It has a long black bill, much curved or arched, about eight fingers only, and begin-

gers from the head. The middle parts of the

the borders or outsides ash-coloured, with an intermixture of red; and those between the wings and back are of a most beautiful glossy blue, and shine like silk. The vent and belly are white. The feet are divided, but joined by a little membrane at the root. The tongue is very short, considering the length of the bill, and bears some resemblance to an arrow.

The female is somewhat larger than the male, which is commonly called the jack curlew, and the spots with which her body is covered almost all over, is more inclining to a red.

THE STONE CURLEW

DIFFERS very much from the former. It is a pretty large bird, being from the extension of the point of each wing a full yard, and has a straight sharp-pointed bill, near two inches long, black towards the nostrils, the other parts yellow; the eyes and the edges of their lids are yellow, there is a bare place under each eye, that appears of a sort of yellowish green, the breast, thighs, and under the chin, are of a yellowish white, the back, head, and neck, are in the middle parts black, with their borders of a sort of reddish ash-colour, with some transverse spots of white upon the quill-feathers, and the outward surface black, some of the other wing-feathers are tipt with white, so that they appear of a fine mixture of black and white, prettily mottled. The tail is about six inches long, the colours variegated like those of the body and wings. The legs are long, and of a yellowish colour, with small black claws; it has only three fore toes, which are joined together by a little membrane: but has not any back toe at all.

They are found in Norfolk, and several other counties in England, and have a cry that very much resembles that of the green plover; they breed very late in the year, insomuch that the young ones have been found in the latter end of October, scarce able to fly; they run very swiftly, and will often stop, and stand without the least motion of any part of their bodies.

THE BARKER.

THIS may not be improperly placed here, from in a great measure partaking of the characteristics of the curlew family, if it does not in reality belong to it. This bird measures from the point of the bill to the end of the tail, near two feet, and from the point of each wing when extended, upwards of three. The head and part of the neck are of a cinereous, or brown colour, interspersed with small black spots; the back, and both the covert and scapular feathers of the wings, of a reddish brown, with white edges and tips: the quill-feathers black, with their outward edges white. The under part of the body is of a dusky white tinctured with yellow. The tail is composed of dusky brown feathers, striped regularly with white on both the webs. The legs

Timidity—Peculiar noise.

and feet are brown, tinctured with a dusky yellow, and greenish gloss.

They generally feed on the salt marshes, not far from the sea, and are so timorous that they will very rarely admit a man to come near them, usually seeking their food in the night as other nocturnal birds do. They are said to make a noise like the barking of a dog, from whence they are supposed to take the name of barker; though, according to Mr. Ray, this appears to be the bird described by Bellonius by the name of berge, and that which the French call petit corlieu, which they esteem a very great delicacy.

THE LARGE WHITE GAULDING

MEASURES from the end of the bill to that of the tail, about three feet and a half, and about four feet from the extension of each wing; the bill is very long, angular, and of a yellow colour, in which there are two long slits for nostrils. The neck is very crooked, resembling in some degree a Roman S, and is about eleven inches long. The feathers that cover the whole body are of an exceedingly beautiful milk-white colour. The thighs, legs, and toes, are about ten inches long, and are covered with large scales, of a bluish black colour. It has four toes, one behind, and three before, the middlemost of which is near three inches long; the claws are black; and

3

there is a small web between the two outermost toes.

It feeds upon small fish, and frequents the sea marshes and salt pools.

Capt. Wood observes, that in the north-west parts of Greenland there is a sort of fowl which the natives catch with springs and snares, chiefly for the sake of their skin and feathers, which being thick, they dress and make garments of them, like furs, wearing the feathers outward in the summer time, and inward in the winter. He says two or three of his men killed fifteen hundred of them in one day.

One would from this account imagine, snares would be as unnecessary here as in the bird island in America, mentioned by the Earl of Cumberland, who says, " there are such incredible numbers of birds found in it, that there needs no artifice to take them ; for a man may catch with his hands alone almost enough to serve a whole fleet."

THE BLUE GAULDING

IS from its bill to the end of the tail about eighteen or twenty inches, and from the ex-

of the bill towards the head is of a bluish colour, and black towards the extremity; it is very sharp, and about two inches and a half long: it has a greenish skin about the eyes, and a tuft of thin small longish feathers upon the head; the neck is about six inches long, covered with thin feathers of a bluish black colour; the whole body of the bird being nearly the same colour, except the breast, belly, and under the wings, which appear something lighter.

The legs are covered with greenish scales, and are about seven or eight inches long; it has four toes, one behind and three before, the middlemost of which is about two inches long; and it has black crooked sharp claws.

They feed on shrimps, young crabs, spiders, and field crickets; and frequent ponds and watery places.

THE BLACK BELLIED DARTER.

THIS species is the size of a common duck; the head, neck, and upper part of the breast are of a pale brown, and on each side of the neck is a broad white line. The belly, wings, and tail are black, as are also the back, scapulars, and wing-coverts; but these are marked with white lines. The bill is bluish above, and somewhat red beneath, and the legs are of a yellowish green. The four toes are united like those of the cormorant. It is about three feet in length, and is found in some parts of Africa, in Ceylon, and Java. In those countries where people are so apprehensive of serpents, whoever sees only foliage of a tree on which it is perched, must naturally mistake it for one of those reptiles, accustomed to climb and reside in trees; and the illusion is increased by its having all the tortuous motions of these animals. In whatever situation it happens to be, whether swimming, flying, or at rest, the most apparent and remarkable part of its body is its long and slender neck, which is constantly in motion, except during flight, when it becomes immoveable and extended, and forms with its tail a perfectly strait and horizontal line.

The principal food of this bird is fish, which,

large it flies off with them to some rock or stump of a tree, and fixing them under one of its feet, tears them to pieces with its bill.

Though water is its principal element, yet it builds its nest, and rears its young on rocks and trees, but always on those that are so near the river, that it can either, in case of danger, or when the young are old enough to learn to swim, precipitate them into it.

There are few birds that exceed these in sagacity and cunning, particularly when surprised on the water. In this situation it is almost impossible to kill them. The head, which is the only part exposed, disappears the instant the flint touches the hammer of the gun; and if once missed, it is in vain to think of approaching them a second time, as they never show themselves more than once, but at a very great distance, and then only for the moment necessary for breathing. In short, so cunning are they that they will often baffle the sportsman by plunging at the distance of a hundred paces above, and rising again to breathe at the distance of more than a thousand below him, and if they have the good fortune to find any reeds, they conceal themselves there, and entirely disappear.

THE WHITE-BELLIED DARTER.

THE white-bellied darter is scarcely so large as a mallard, but its neck is so long that it measures not less than two feet ten inches. The bill is three inches long, straight, and pointed. The neck is covered with downy soft feathers, of a reddish grey; the upper parts of the plumage are dusky black, dashed with white; the under parts pure silvery white. It is a native of Brasil, and is extremely expert at catching fish.

Mr. Bertram, in his American Travels, says that these birds have a way of spreading out their tail like an unfurled fan. They delight to sit in little peaceable communities, on the dry limbs of trees, hanging over the still waters, with the wings and tail expanded, and when approached, they drop from the limb into the water, as if dead, and for a minute or two are not seen, when on a sudden, at a vast distance, their long heads and necks are raised, and have much the appearance of snakes, as no other parts of the body are to be seen when swimming, except sometimes the tip of the tail. In the heat of the day they are often seen in great numbers, sailing high in the air over the rivers and lakes.

THE BOATBILL,

A VERY curious bird, is found in the southern parts of America, of which it is a native; it is about the size of a common fowl. The general colour of the bill is dusky, and the skin beneath the under jaw is capable of distention. From behind the head springs a long black crest. The plumage on the forehead is white, and the rest of the bird is a pale bluish ash colour; and the feathers which hang over the breast are loose, like those of the heron. There are varieties of this bird, both spotted and brown, but they appear simple varieties, and not at all enti-

king-fisher, it preys upon fish, which it catches by perching on trees that over-hang the streams, and dropping on the fish as they swim by.

THE UMBRE

TAKES its name from its colour, which is of a deep brown, or umbre, and is brought from the Cape of Good Hope; it is about the size of a crow, to which it is very similar. The bill is three inches and a half in length, with a furrow on each side the upper mandible, and from the head springs a large crest of black feathers, better than four inches in length.

THE JACANA

IS found in most of the tropical climates, but is most common in South America. It is remarkable for the length of its toes, and for the wings being armed in front with sharp spurs. There are about ten species, differing in size from that of a common fowl to that of a water-rail. They vary also in their plumage, some brown, some black, and some variable. The faithful jacana is a most useful bird-at Carthagena in South America. The natives, who keep poultry in great numbers, have one of these tame, who attends the flock as a shepherd, to defend them from birds of prey. Though not larger than a dunghill-cock, the jacana is able, by means of the spurs on his wings, to keep off birds as large as the carrion vulture; and even that bird himself; and it never deserts its charge, but assiduously takes care to bring the whole flock safe home at night. It feeds on vegetables, and cannot run but by the help of its wings.

THE KOKOÍ,

A BRASILIAN bird of the crane kind, and very pleasing to the sight, is about the size of a stork; its bill is straight and sharp, about six fingers in length, of a yellowish colour, inclining to green; the neck is fifteen fingers long, the body ten, the tail five; the neck and throat are white; both sides of the head black, mixed with ash-colour. On the undermost part of the neck, are most delicious white, long, and thin feathers, fit for plumes; the wings and tail are of an ash-colour, mixed with some white feathers; all along the back, are long and light feathers like those on the neck, but of an ash-colour; the legs are very long, and covered above half the way down with feathers. Its flesh is very good, and of a grateful taste.

Such are the several aquatic birds known to naturalists; but there is no doubt, from the many discoveries which are continually made by navigators and travellers, that there are various others, not only of water, but land birds hitherto unknown to mankind: it is likewise very probable that some of our voyagers have mistaken different species of known birds, for some unknown genera. We shall now conclude our descriptions with some observations on

SHOOTING WILD-FOWL.

" TO be equipped for this sport," observes the Rev. Mr. Daniel, " in severe weather it is essentially requisite to be well clothed."—Dr. Harward, one of the best wild-fowl shooters in the kingdom, recommends, if a sportsman's boots be new, that they should be well anointed with half a pound of bee's wax, a quarter of a pound of resin, and the like quantity of mutton suet, or tallow, boiled together, and used lukewarm. Should the boots have been used, beef suet is to be substituted for the mutton. Thus the boots are rendered water-proof; under which the sportsman should wear thick yarn stockings, and over them what is termed wads by the fishermen. A cap must be worn, made of skin, instead of a hat: the fowl will not approach near the latter, and nothing so much or so soon shies them.

The punt-shooters (men who earn their livelihood by attacking the wild-fowl night and day, according as the tide serves) kill great numbers.

The pursuit is hazardous, especially when there is much ice in the river, by which they some-times get encircled, and then can only float with the current, and are kept often two or three tides before they can extricate themselves, and their punt is ill calculated to sustain pressure against its tides, which are not twenty inches high from the surface of the water; in this the punter by night drops down with the tide, or uses his pad-dles after the fowl: he knows their haunts, and takes every advantage of wind, tide, moon, &c. His gun, which carries as much as a little can-non, is laid with the muzzle over the stem of the punt, in a hitch which regulates the line of aim: at the bottom of the punt he lies upon his belly, and gets as near the rout of fowl that are upon the water as possible. When within the range of his gun, he rattles with his feet against the bottom of his punt, and when the fowl begin to spring at this unexpected sound, at that moment he pulls the trigger and cuts a lane through their ranks; he instantly follows the direction of his shot and gathers up those that are killed, or just expiring, for very seldom he makes it answer to row after fowl only wounded; he then charges his gun and drifts further down the river, in hopes of a second, third, and successive shots. By this mode one man has brought home from four-score to an hundred wild-fowl, of various kinds in one night's excursion.

The best time for this shooting is the first or second day's thaw after a sharp frost, and when deep snow has long covered the ground: the fowl are then seen flying in every direction to dabble in the fresh-water, which then appears all around inviting them. Another favourable opportunity is at the commencement of a frost, with the wind strong at east, and a sleet or snow falling; if the guns can but be kept dry, there is no complaint about the using them, and the fowl in such weather always fly lower than when the atmosphere is clear.

The gun proper for this amusement has no occasion to be more than three feet eight inches in the barrel, which should not weigh less than twelve pounds; upon this scale the whole gun will be about eighteen pounds weight; this quantity of iron at the above length will be as capable or more so, of throwing shot sharp and distant, as a barrel two feet longer. Should this heavy mass be objected to as cumbersome to carry, let it be remembered that these guns are not meant to lie upon the arm, or to be carried about in the fields. The shooter is either seated in a boat, or upon a marsh; in either situation the gun does not fatigue him, since he has nothing to do but elevate it as the wild-fowl fly over his head; and after firing and charging, let it again lie beside him until fresh objects require its use.　　　　　2

For keeping a sportsman's gun clean, Dr. Harward recommends three ounces of black lead, half a pound of hog's lard, and one quarter of camphor, boiled upon a slow fire. The gun barrels are to be rubbed with this preparation, and after three days, wiped off with a linen cloth. **Twice in a winter will keep off the rust, which** the salt-water is otherwise sure to be continually bringing out of the iron.

END OF VOL. IV.

INDEX.

A

B

D

H

M

N

O